INDUSTRIAL
HYDRAULICS

INDUSTRIAL HYDRAULICS

Richard W. Vockroth, Ph.D.

Delmar Publishers Inc.

I T P™

NOTICE TO THE READER

Cover design by Lisa Bower

Delmar staff:
Publisher: Michael McDermott
Administrative Editor: John Anderson
Development Editor: Sheila Davitt
Project Editor: Barbara Riedell
Senior Production Supervisor: Larry Main
Art and Design Coordinator: Lisa Bower

For information, address: Delmar Publishers Inc.
3 Columbia Circle, Box 15015, Albany, NY 12203–5015

COPYRIGHT © 1994 BY DELMAR PUBLISHERS INC.

The trademark ITP is used under license.

Printed in the United States of America
Published simultaneously in Canada
by Nelson Canada,
a division of The Thomson Corporation

ISBN 0-8273-5644-7

4   5   6   7   8   9   10   XXX   01

Library of Congress Cataloging-in-Publication Data

Vockroth, Richard William, 1931––
   Industrial Hydraulics / Richard W. Vockroth.
      p.   cm.
   Includes index.
   ISBN 0-8273-5644-7
   1. Hydraulic machinery   2. Hydraulic engineering.   I. Title.
TJ840.V63   1993
620.1′06 – dc20
                                        93–34565
                                    CIP

Contents

Preface

Societal evolution and rapid advances in technology have forced dramatic changes in vocational–technical and engineering education since the midpoint of the 20th century. Awareness and opportunities have raised the aspirations of those who might otherwise have settled for a routine livelihood. Recurrent learning is now an integral part of employment, especially in technical disciplines. Increasingly, tasks once reserved for holders of baccalaureate degrees are assigned to specialists with lesser credentials. Many topics which were covered in freshman college courses are introduced in vocational secondary schools.

These developments create major problems for vocational–technical educators. While technology becomes increasingly complex, they are obliged to deal with students lacking key prerequisite skills, particularly in communications and mathematics. Textbook authors persist in assuming cognitive levels not always met by readers. Teachers resort to manuals written by manufacturers, intended for maintenance and repair technicians rather than for beginning students.

Introduction to Industrial Hydraulics is intended primarily, but not exclusively, for prospective learners whose needs have been neglected. A frequent comment of reviewers, all postsecondary educators, was that the material is "too elementary." Not one, however, criticized it for lack of clarity. This is as it should be.

Several strategies and devices have been incorporated specifically to maximize reader comprehension:

- Principles of fluid behavior, force, pressure, flow, volume, and capacity, which form the foundation for all that follows, are introduced early;

- An entire chapter comprises an introduction to the basic hydraulic system, showing the part each component plays in system operation;
- Mathematical concepts are presented as needed, so algebra need not be a prerequisite;
- Many problems, ranging widely in level of difficulty, follow each chapter. The instructor can select those appropriate to the class level;
- Key concepts and definitions are isolated, **boldfaced**, and indented for emphasis and to facilitate later review;
- Key words are **boldfaced** for emphasis, relieving the learner of the need for highlighting while studying;
- Explanations are repeated when appearing within more than one context, for reader convenience;
- Vocabulary is kept at the lowest reasonable level. Unfamiliar terms are defined both in the text where first used and in the glossary;
- Wherever a sequence of operating motions needs to be shown, individual illustrations show successive positions;
- All line drawings were created specifically for this text, showing principles of operation. Typically, textbooks rely heavily upon illustrations from maintenance manuals intended for other purposes.

For a textbook of this nature, the real measure of worth is the extent to which the reader, having mastered its contents, can perform in the workplace. This author accepts that challenge.

I wish to express appreciation to my colleague, Professor James "Ted" Horigan of Corning Community College for advice and criticism during the preparation of the manuscript.

Learning about Hydraulics

1

In an age when technical education seems to be all about the newest developments in electronics, computers, and automation, you may wonder whether it is really worthwhile to study a technology that has been around for centuries. Do we really need hydraulics any more? Are you wasting your time and effort mastering a field that may be obsolete in a few years?

Virtually all industrial processes are based upon the use of energy to do work in just four types of systems—**mechanical, electrical, hydraulic,** and **pneumatic.** If you examine the innovations and new developments in products and methods, you will find they represent new ways of **applying** and **controlling** these systems. Knowledge and understanding of the underlying principles upon which mechanical, electrical, hydraulic, and pneumatic systems are based are every bit as important as ever. Yes, you are about to learn the basics of a very important and valuable technology—the applied science of hydraulics.

HISTORICAL PERSPECTIVE

It is likely that primitive man's first use of natural assistance to do work involved various combinations of gravity, the principles of leverage, and inertia. A rock dropped from some height produces more force than its own weight, and could be used to break a branch, kill an animal, or make a hole in the ice covering a lake. The leverage developed by a pole wedged under a rock or tree root increases the applied force by many times. As a result of its **inertia,** the impact of a swinging club is far greater than from one that is merely pushed. Likewise, the harder a ball is thrown, the more it hurts when caught; it also takes less energy to keep a cart rolling once it is moving than to get it started.

> **Inertia is defined as a property of matter by which it will remain at rest or in motion in a straight line until acted upon by an external force.**

Of course the cavemen had no dictionaries. People were using mechanical principles thousands of years before they were analyzed, put into mathematical formulas, and taught in schools and colleges.

Eventually people realized that air and water also could be harnessed to serve useful purposes. As Figure 1-1 illustrates, sails were raised into the wind to move large ships. Windmills were built to draw water from wells and irrigate cropland. Water wheels were erected beside streams and coupled with mechanisms to process grain. Prior to the discovery of electricity, air and water were the only providers of a continuous flow of energy capable of doing work.

We now use the term **fluid power** to describe schemes using air and water to do this, although as we shall see, our modern systems are quite different.

Figure 1-1 Fluid power has served mankind for centuries

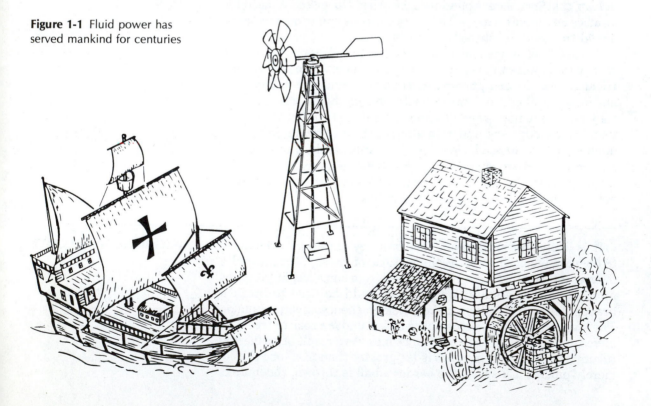

Fluid Power is an applied science dealing with ways of using pressurized gas or liquid for doing work.

When a gas (usually air) is used, we have a **pneumatic** system. When a liquid (usually oil) is used, we have a **hydraulic** system. Pneumatics and hydraulics are the two forms of **fluid power**, and the characteristics of the two kinds of systems are quite different. Some devices and applications take advantage of the special qualities of both by using them in combination. The well-rounded technician or engineer must understand mechanical, electrical, hydraulic, and pneumatic principles to choose and apply them properly.

Every substance on earth exists in one of three forms — **solid**, **liquid**, or **gas.** Most substances are capable of **changing** form, as when steel is melted, or when ice is heated to become water and then boiled to become water vapor, a gas.

The differences are quite clear: (1) A **solid** has a specific **volume** (size) and **shape**, and neither of these is affected by the shape of a container; (2) a **liquid** has a specific **volume**, but will take on the shape of its container and will change if it is put into a container having a different shape; and (3) a **gas** has **no** specific **shape or volume** and will fill the shape and size of its container. Figure 1-2 illustrates these properties.

Mechanical systems use solid materials such as metal, wood, or plastic formed into shafts, gears, pulleys, levers, rods, cams, and other parts to transmit power. **Hydraulic** systems use a liquid,

SYSTEM CHARACTERISTICS

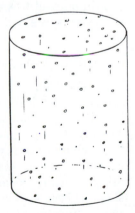

SOLID
MAINTAINS BOTH ITS SHAPE AND ITS VOLUME.

LIQUID
MAINTAINS ITS VOLUME AND TAKES THE SHAPE OF ITS CONTAINER.

GAS
TAKES ON BOTH SHAPE AND VOLUME OF ITS CONTAINER.

Figure 1-2 Characteristics of a solid, liquid, and gas

usually oil, which is pumped into hydraulic cylinders or hydraulic motors and made to do work. **Pneumatic** systems use a gas, usually air, which is pressurized and then directed into pneumatic cylinders and made to do work. **Electrical** systems use energy in the form of electron flow to cause mechanical motion, to heat or cool materials, or to remove material in manufacturing processes.

Early manufacturing plants were totally mechanical. Steam engines, connected to long shafts, provided the energy. These shafts were hung from the ceiling and used to power machines throughout the plants by leather belts. Power transmission was accomplished entirely by **mechanical** means. A trained person was assigned to each machine, and the quality of what was produced varied according to the skill of the operator. The **control** of the machines and processes was dependent upon the talent, experience, and care of this individual.

As you know, electrical power replaced the steam engines, belts, and shafts, and the machines are now driven by electric motors. Electronic controls and computers have taken over many of the tasks involved in running the machines. It is important to remember, however, that the machines are still mechanical; only the ways in which they are **powered** and **controlled** have changed.

SYSTEM COMPARISONS

Nearly all actual operating systems in industry make use of **combinations** of mechanical, electrical, hydraulic, and pneumatic principles. Milling machines, lathes, and drill presses are driven by electric motors. Robot arms with grippers often consist of mechanical linkages moved by pneumatic cylinders. Hydraulic cylinders are used to form plastic parts for automobile interiors. It is important to know the strengths and limitations of the types of systems available.

Advantages of Mechanical Systems

Because they have been around longer than others, more is known about mechanical systems, and they tend to be more reliable. There are mechanical devices centuries old that are still working. Thousands of parts such as gears, shafts, levers, and pulleys have been standardized and are readily available for replacement when needed. Mechanical systems pose little or no safety hazard when not operating; no pressurized lines to burst or electric lines to shock. These systems are generally simpler to troubleshoot and repair since broken or damaged parts can be readily found and replaced. Finally, with reasonable precautions, mechanical systems and devices tend to be more reliable than other types in harsh environments such as high heat, extreme cold, dust, chemicals, abrasives, or radiation.

Weaknesses of Mechanical Systems

The material which provides strength, such as iron, steel, or brass, is also heavy and adds inertia, slowing the system's response to changes in motion. Moving parts need periodic lubrication and protection from oxidation and corrosion, as well as repair or replacement as they wear or become damaged. The weight and inertia of the material in rods, chains, belts, gears, and other parts limit the distance over which power can be transmitted economically and effectively. Finally, every part in the system must be made strong enough to carry the largest load for which the system is designed, even if the usual load carried may be much less. The weight and inertia of all these parts must be overcome whenever the system operates.

Advantages of Electrical Systems

High voltage power lines provide the most convenient means of transmitting power over long distances with low power loss. There is no weight or inertia in electron flow. A single electric power generating plant provides energy for many factories and individual machines, increasing efficiency and lowering cost. Machines and motors ranging in size from less than a horsepower to hundreds of horsepower can operate efficiently from the same power source. The amount of energy used can increase or decrease with the needs of the industry, and the user need pay only for what is used.

Weaknesses of Electrical Systems

The potential for electrocution or electrical fire is ever present, whether the system is operating or not. The system is dependent upon an outside source (the electric company) for a continuous supply of energy. Electrical energy, in the large amounts required by industrial processes, cannot be stored and must be used as it is delivered. To be efficient, electric motors must be run at relatively high speeds and must usually be reduced by mechanical means which consume energy and are wasteful. The range of speed control, even with DC (direct current) motors, is small and the efficiency of these motors at slow speed is low.

Advantages of Hydraulic Systems

Force levels can be made very high, just by using large diameter cylinders. This force can be transmitted over moderate distances at uniform intensity and through small passageways and around corners. The amount of force or motion can be controlled with a high degree of precision. Motion can be reversed very quickly because there is little inertia, and it can be started and stopped instantly with no slack or backlash unlike when belts, chains, or gears are used. Hydraulic systems can be assembled into self-contained, easily maintained, portable units, as are found on earth-moving equipment such as bulldozers, backhoes, graders, and power shovels. By using an accumulator, (an energy storage device), a hydraulic system can be made to deliver more horse-power for short periods than that provided by the motor driving the system, and it can store energy for reserve or emergency use.

Weaknesses of Hydraulic Systems

A hydraulic system must be totally enclosed with a system of lines, valves, cylinders, and a tank, and there is nearly always some leakage, requiring constant maintenance and clean-up. The entire system must be located close to the point of actual usage, or pressure drop and energy waste may become excessive. There is always a fire hazard if petroleum oil is used, and fire-resistant fluids are expensive. If a pressurized line should burst, people could be injured or killed, and equipment could be damaged or destroyed. Speed of operation is limited because if the hydraulic oil flows too fast, it becomes turbulent and loses pressure. Finally, hydraulic systems require a carefully planned and executed system of maintenance, and there is a serious shortage of people trained in this specialty.

Advantages of Pneumatic Systems

Since air is readily available, it can be taken into the system, used, and vented to the atmosphere with no need for a tank or return line. Pressure loss in air lines is very low, so a single com-pressor can supply pressurized air to several machines placed a great distance away within a factory. Motion can be reversed very quickly because there is little inertia, and force can be easily transmitted through small passageways and around corners. By using a receiver (enclosed airtight tank), a pneumatic system can store energy for reserve or emergency use. Pneumatic devices typically respond very quickly when activated.

Weaknesses of Pneumatic Systems

Because any gas is resilient, or "springy," precise speed or posi-tion control cannot be accomplished readily. When air is com-pressed, heat is produced requiring cooling and wasting energy. Air taken directly from the atmosphere needs to be filtered and dried, and lubricant must be added by devices called "conditioners" before it is used. These require constant maintenance. Leaks in a pneumatic system can produce a very high pitched noise, which is hazardous to hearing. Finally, pneumatic systems require a carefully planned and executed system of maintenance, and there is a serious shortage of people trained in this specialty.

This is only a brief comparison of the systems available, but as you can see, each has its own special benefits along with its disadvantages.

<hr />

EVOLUTION OF FLUID POWER

From an engineering standpoint, those working in the fields of hydraulics or pneumatics are concerned with either **transporting** a fluid from one place to another or **using** a fluid to do work. We still find remains of aqueducts, canals, tunnels, and piping throughout the world dating back to ancient times. These systems carried water for drinking, washing, floating vessels, and related purposes. In such cases, it was the **water itself** that was needed, not its ability to transmit and use energy.

The first applications in which fluid was used to do work were all what we now call **open** systems because the air or water used was not confined, and it was **weight** or **inertia** that provided the force needed. Examples still in use include windmills, sails, water wheels, and even giant hydroelectric power plants. These all require **large amounts** of **moving fluid** at **low pressures** to accomplish their purpose.

Historians credit Englishman Joseph Bramah with inventing the **closed** fluid power system, when he assembled the first hydraulic jack, or press, in the late 1700s. As Figure 1-3 illustrates, Bramah

Figure 1-3 Operating principle of Bramah's jack

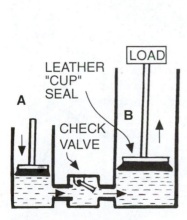

(1) Piston **A** is pushed down. Check Valve opens. allowing water to flow. Piston **B** lifts load.

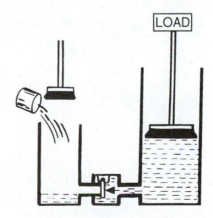

(2) Piston **A** is raised, and water is added. Check valve blocks return flow. Piston **B** supports load.

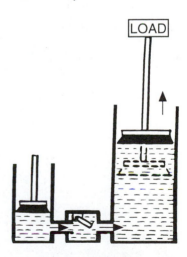

(3) Piston **A** is pushed down again. Check valve opens, allowing flow. Piston **B** lifts load higher

pumped water into a cylinder through a special valve called a check valve, which allowed flow to the cylinder but blocked it from returning. By repeating this operation several times, he made a piston rod extend and exert force. This is the same principle used today in hand-operated hydraulic jacks and piston pumps. While simple in construction, it was the first practical hydraulic device to increase an applied force using pressure higher than that provided naturally. Figure 1-4 shows one example of a hydraulic hand pump.

Fluid power as an energy source gained wide acceptance. By the mid-1800s, several large cities in England were using central water pumping stations and a piping network to supply power to factories. This continued until electric motors became popular, in about 1900. Electric power was found to be much more efficient and convenient to deliver over great distances, and this continues to be the way industry is powered today. Wherever fluid power systems are used in manufacturing, you will usually find them assembled in compact units with electric motors driving their pumps or compressors (see Figure 1-5).

Engineers have long understood that each design problem has its own particular requirements, and electric power does not always offer the best solution. When the battleship USS *Virginia* was launched in 1906, its guns were moved and controlled by means of fluid power. This technology was found to be the best

Figure 1-4 The principles developed by Bramah are used today in hydraulic hand pumps. *Courtesy of Power Team Div., SPX Corp., Owatonna, MN*

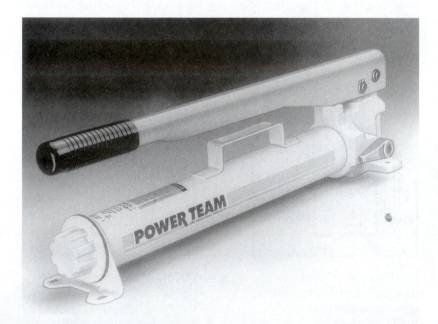

means of providing the force and accuracy needed. By this time, it was realized that water has serious disadvantages as a hydraulic fluid:

1. Water evaporates, especially when heated;
2. Water freezes;
3. Water, being a "thin" (low viscosity) liquid, leaks readily from the system and also leaks past the piston seals inside cylinders;
4. Water rusts parts in the system;
5. Water is a poor lubricant, allowing unnecessary friction between moving parts.

Figure 1-5 Using a "building block" concept, hydraulic systems can be assembled to be powered by gasoline, electricity, or compressed air. *Courtesy of Power Team Div., SPX Corp., Owatonna, MN*

Figure 1-6 Hydraulic portable crane. *Courtesy of Air Technical Industries, Mentor, OH*

Natural petroleum oil was found to have none of these drawbacks. For these reasons, oil was used instead of water as the fluid on the USS *Virginia* and even today remains the popular choice for most hydraulic systems. Fluid power has become the favored means of powering much shipboard equipment by today's navies throughout the world.

The type of system described in Chapter 4, in which the various parts of a circuit are assembled as a "package," or a "unit," was first developed in the United States in the 1920s. Although various improvements have been made over the years, it remains basically unchanged. Typically, power to drive the pump is provided by a gas or diesel engine, or an electric motor. This is located close to where the forces and motion are needed in order to reduce energy losses due to friction in the piping. One example is the hydraulic portable crane, shown in Figure 1-6.

CAREERS IN FLUID POWER

In the fluid power field, as with any technology, there are many skill levels involved ranging from the operation of simple tools driven hydraulically or pneumatically, to the design and building of very complicated machine tools and controls like those in Figures 1-6 and 1-7. Unfortunately, because of the shortage of trained people, equipment is often operated by individuals with little under-

Figure 1-7 Fluid power is used extensively in the operation of earth-moving equipment. *Photo by the author*

standing of how or why it works. This leads to safety hazards, inefficiency, equipment breakage, excessive maintenance expense, and poor quality in the work being produced. As you grow in knowledge and experience, you become more valuable as an employee and find yourself being given more interesting and challenging work, and of course, higher pay. Let's look at a few examples of occupations in fluid power.

Hydraulic Equipment Operators set up, operate, and perform routine maintenance tasks on machines and equipment involved in such activities as producing or assembling parts, moving products or equipment, constructing buildings and bridges, and harvesting crops. Training at this level is relatively short, usually a matter of a few weeks. Pay is generally low, since operators can be easily replaced.

Equipment Technicians install, repair, make major changes and improvements when needed, and perform skilled maintenance on all types of hydraulic equipment. Training at this level normally takes at least several months, and many technicians hold two-year college degrees. They are often sent to specialized schools for further training on complex equipment such as robots or automated machines like those shown in Figures 1-8 and 1-9. Trained, experienced technicians earn good pay and, equally important, find their work interesting and challenging.

Sales Representatives travel to factories, laboratories, and job sites to meet with customers, equipment designers, technicians, and engineers to help them solve problems and write orders for equipment. To do this well, they need training at least at the level of technicians, plus experience with several different kinds of equipment. In addition, they must be very knowledgeable about

Figure 1-8 Hydraulic puller capable of forces up to 55 tons. *Courtesy of Power Team Div., SPX Corp., Owatonna, MN*

Figure 1-9 Forty 55-ton rams were used to raise 1.4 million pounds of bridge to meet new interstate highway standards for overpass clearance. *Courtesy of Power Team Div., SPX Corp., Owatonna, MN*

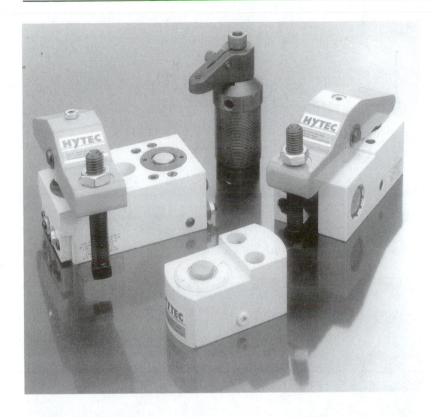

Figure 1-10 Hydraulic work-holding devices permit rapid clamp/release of parts in automated manufacturing systems. *Courtesy of Power Team Div., SPX Corp., Owatonna, MN*

the products their employer sells and be able to work well with others. Their pay ranges from good to excellent, depending upon their knowledge of the product line, and their ambition, initiative, and ability to work well with others.

Field Engineers visit factories, laboratories, and job sites where hydraulic equipment is installed. They supervise, install, modify, or troubleshoot the equipment. They are called in as the "experts" when problems prove too difficult for the customer's technicians and maintenance personnel. You normally earn this title with several years of experience in solving practical problems. The pay ranges from good to excellent, but equally important is the feeling of pride in knowing that you are the one called to solve problems when others have failed.

System Designers work in offices or laboratories using drafting boards or Computer-Aided Drafting (CAD) equipment to lay out hydraulic circuits. Given information about what needs to be accomplished hydraulically, they select pumps, valves, cylinders, and other devices as needed to do the job. The drawings they produce are then used to assemble the circuits. Designers must have a thorough understanding of all kinds of hydraulic devices such as those in Figure 1-10 and the ability to invent new ways of using

Figure 1-11 Hydraulic principles can be adapted to a wide variety of presses. *Courtesy of Multipress Div., Quality Products Inc., Columbus, OH*

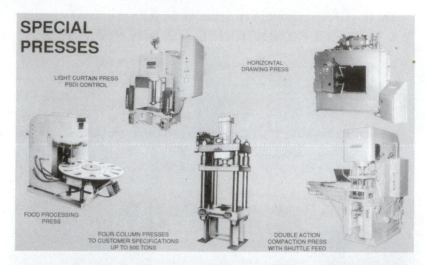

Figure 1-12 Hydraulically operated flight simulator. *Courtesy of Parker-Hannifin Corp., Fluidpower Group*

them. They must be able to look at symbols on drawings and imagine how the entire circuit will operate when put together. To be employed at this level, you will probably need a college degree in engineering, and the pay is generally high.

These are only a few of the many kinds of positions you might obtain in the field of hydraulics. In this, as in all technical occupations, you need study, training, and experience for all but the simplest and lowest-paying jobs. As you advance, you will likely be getting further training, your earnings will increase steadily, your work will become more interesting and challenging, and your friends and co-workers will respect you more for what you have accomplished.

SOLVING TECHNICAL PROBLEMS

People have difficulty in solving technical problems because they do not begin with an **organized approach**. This means following a plan which begins with writing a mathematical statement (called a **formula**) about the problem, then replacing unknown values in the formula with the actual figures at hand. This is done with a series of **equations**, substituting and simplifying values until you have the answer to the problem.

An equation is a statement that two values, or groups of values, are equal.

One example of an equation is

$$7 + 2 = 3 \times 3.$$

The sum of the figures on the left of the equal sign is the same as the answer you get by multiplying the figures on the right. In this example, nothing is "solved" because all the numbers are given.

Suppose, however, we are told that the number of pencils in a box is the same as 3×3. This too could be written as an equation, but rather than write so many words, we can just let n stand for the *n*umber of pencils and write the equation as

$$n = 3 \times 3.$$

Multiply to find the answer, indicate what you have found, and label it as the solution to the problem:

$$n = 9 \text{ pencils (Ans.).}$$

The reason for labeling is that we often have problems in which we have to solve for one or more values to get to the answer to the original question. It is only this final answer that is labeled. The steps in solving technical problems are:

1. Read the problem carefully. Determine exactly **what you are being asked to find,** and what information is given to help you in your solution;
2. Decide **how you will use** the information given. For some problems you need only choose a formula you have learned and substitute numbers for the letters. For others, you may have to do some arithmetic first to find the numbers required by the formula;
3. Choose the letters you will use to represent the **unknown** values and write them down. It is helpful to choose letters that suggest what they represent, and write down what they mean. For example:

$$D = \text{Diameter}$$
$$L = \text{Length}$$
$$H = \text{Height;}$$

4. Write the problem as an **equation;**
5. Solve the problem **in steps,** doing the arithmetic needed until you have the answer;
6. Write the answer in the **units** required, such as inches or pounds, and write "(Ans.)" after it to indicate that this is **your answer** to the original problem;
7. Go back and **reread** the original problem, to make sure that you have answered the question asked.

Now let's solve a problem, applying these seven steps.

Problem: You need 6 hydraulic cylinders for a project. They will cost $70 each, and the total shipping cost is $30. How much is the total bill?

1. You are to find the total cost of the order (6 cylinders) plus the cost of delivery (shipping). You are given the cost of one cylinder and the shipping cost, which is the same regardless of the number of cylinders ordered;
2. This will be a two-step problem. First, find the total cost of the 6 cylinders. Then add the shipping cost. This will be the answer;
3. Let t represent the total bill, let c represent the total cost of the cylinders, and let s represent the shipping cost;
4. The total bill is the cost of the 6 cylinders plus shipping, so the equation is:

$$t = c + s;$$

5. Now solve the problem in steps. First find c, the cost of the 6 cylinders.

$$c = 6 \times \$70$$
$$c = \$420$$

Now use this in the equation from Step 4:

$$t = c + s$$
$$t = 420 + 30$$
$$t = 450;$$

6. The cost is in dollars, so the last step should be written as

$$t = \$450 \text{ (Ans.)};$$

7. A re-check of the original question shows that this is what was asked for—the total cost of 6 cylinders plus shipping.

CHAPTER SUMMARY

Most industrial processes involve the use of energy to do work in just four types of systems—**mechanical, electrical, hydraulic,** and **pneumatic.** Recent innovations and developments represent new ways of **applying** and **controlling** these systems.

Primitive man's first use of natural assistance to do work was by means of **mechanical** devices, taking advantage of **gravity, principles of leverage,** and **inertia.** Prior to the discovery of electricity, air and water were the only providers of a **continuous flow** of energy available to do work. **Fluid power** is an applied science dealing with ways of using pressurized gas or liquid for doing work. **Pneumatic** systems use a gas, usually air, while **hydraulic** systems use a liquid, usually oil. Many practical devices use the two in combination. Every substance on earth exists as either a **solid,** a **liquid,** or a **gas.** Most substances are capable of existing in more than one form.

Mechanical, electrical, hydraulic, and pneumatic systems all have their individual advantages and weaknesses. It is important for anyone working in engineering or technology to be aware of these. Fluid Power offers a wide range of career opportunities at several levels and requires a variety of personal skills and interests.

The main reason many people have difficulty in solving technical problems is that they do not begin with an **organized approach.** This means setting up and following a **logical plan.** This means selecting one or more formulas, substituting given values for the letters or symbols, and performing the mathematical operations.

PROBLEMS

1.1 Name the four basic types of systems used in industry to do work.

1.2 Explain the difference between hydraulics and pneumatics.

1.3 Define "inertia" and describe four examples in which it is used to help us do work.

1.4 List four advantages and four weaknesses of mechanical systems.

1.5 List four advantages and four weaknesses of electrical systems.

1.6 List four advantages and four weaknesses of hydraulic systems.

1.7 List four advantages and four weaknesses of pneumatic systems.

1.8 Explain what we mean by the term, "fluid power."

1.9 What are the three forms in which all substances exist, and how do they differ from each other?

1.10 What kinds of engines were used in factories before electric motors were invented?

1.11 List three reasons why water is not usually used in industrial hydraulic systems.

1.12 Describe briefly what is meant by the term, "open fluid power system," and list three examples.

1.13 Who, according to historians, invented the first practical "closed" hydraulic system, and what fluid was used?

1.14 Write, as an equation, this statement: "The number of pencils in my desk equals the number of yellow pencils plus the number of red pencils plus the number of blue pencils." Use n to represent the total of all the pencils, y the number of yellow pencils, r the number of red pencils, and b the number of blue pencils.

1.15 Write, as an equation, this statement: "The number of vehicles in a parking garage equals the number of cars plus the number of motorcycles plus the number of trucks." Use v to represent the total number of vehicles, c the number of cars, m the number of motorcycles, and t the number of trucks.

1.16 One gallon of hydraulic oil weighs 7.5 lbs. Write an equation for finding the total weight (represented by w) of any number of gallons of hydraulic oil (represented by n).
(a) Use this equation to find the total weight of 8 gallons.
(b) Use this equation to find the total weight of 20 gallons.

1.17 One gallon of hydraulic oil weighs 7.5 lbs. One gallon of water weighs 8.3 lbs. Write an equation for finding the total weight (represented by t) of h gallons of hydraulic oil plus w gallons of water.
(a) Use this equation to find the total weight of 3 gallons of hydraulic oil plus 2 gallons of water.
(b) Use this equation to find the total weight of 1 gallon of hydraulic oil plus 3 gallons of water.

1.18 You have a system whose hydraulic pump delivers 1.5 gallons of oil per minute. Write an equation for finding the total number of gallons delivered (represented by g) in any specified number of minutes (represented by m).

    (a) Use this equation to find the number of gallons delivered in 9 minutes.

    (b) Use this equation to find the number of gallons delivered in one hour.

1.19 One gallon of hydraulic oil weighs 7.5 lbs. One gallon of water weighs 8.3 lbs. Write an equation for finding the difference in weight (represented by d) between h gallons of hydraulic oil and w gallons of water.

    (a) Use this equation to find the difference in weight between 5 gallons of hydraulic oil and 3 gallons of water.

    (b) Use this equation to find the difference in weight between 7 gallons of hydraulic oil and 2 gallons of water.

Area, Force, and Pressure

2

The basis upon which all hydraulic systems operate is the relationship between **area, force,** and **pressure.** It is the ability of a fluid to distribute an applied force to one or several locations and to control the amount of force at each location that makes this technology unique. The many applications, ranging from a simple bottle jack to hydraulic automobile brakes to huge industrial presses, represent, in the final analysis, ways in which engineers have related area, force, and pressure in fluid systems to practical purposes.

Area is the measure of a flat surface, and regardless of its shape is given in "square" units such as square inches or square millimeters (see Figure 2-1).

AREA

Figure 2-1 Each shape has an area of 4 square inches

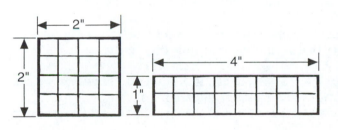
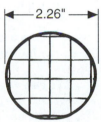
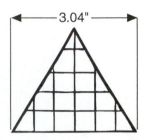

To find the area of a rectangle, multiply its length by its width using the formula

$$A = L \times W$$

where

A = Area to be found
L = Length
W = Width.

Example: Find the area of a rectangle which measures 9.5″ × 8.0″

$A = L \times W$
$A = 9.5 \times 8.0$
$A = 76$ sq. in. (Ans.)

We can better express this formula using a method taught in algebra classes:

To show that two numbers are to be multiplied, we enclose one or both of them in parentheses.

For example, we can write 3×7 as $3(7)$ or as $(3)(7)$. The following method is also commonly used:

In a formula, where we use letters for values which will later be replaced by actual numbers, we can omit the parentheses.

We can write $L \times W$ as $L(W)$ or $(L)(W)$ or simply LW.
Since all four sides of a square are the same, we let s represent the length of a side in the formula. Instead of writing

$$A = (s)(s)$$

we use another method taught in algebra classes:

Place a small "2" above and to the right of a letter or number to indicate that it is to be multiplied by itself.

We can write $s \times s$ as s^2 and call it "s squared." Likewise, we can write 7×7 as 7^2 and call it "7 squared."
Now use the shortened form of the formula to find the area of a square

$$A = s^2$$

where

$$A = \text{Area to be found}$$
$$s = \text{length of a side.}$$

Example: Find the area of a square which measures 5″ per side.

$$A = s^2$$
$$A = 5^2$$
$$A = 25 \text{ sq. in. (Ans.)}$$

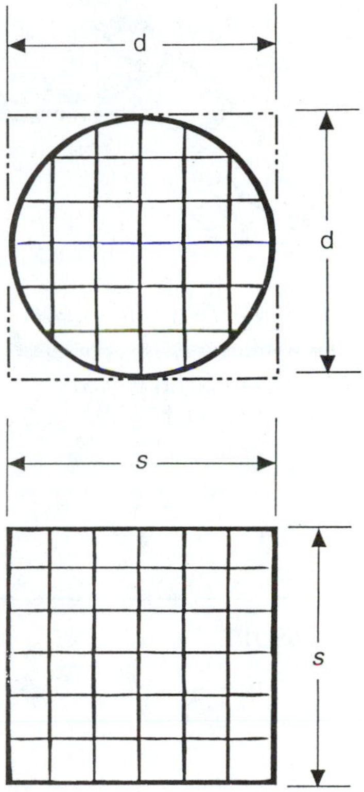

If we compare a circle and a square having the same measurement as in Figure 2-2, it is obvious that the circle has less area. We could find the area of a circle using the formula

$$A = \pi r^2$$

which you may have learned. Since hydraulics problems usually are concerned with cylinders for which the diameter rather than the radius is given, we use a simpler formula

$$A = .785d^2$$

where

$$A = \text{Area of the circle}$$
$$d = \text{diameter given.}$$

Example: Find the area of a circle having a diameter of 2″.

$$A = .785d^2$$
$$A = .785(2)^2$$
$$A = .785(4)$$
$$A = 3.14 \text{ sq. in. (Ans.)}$$

You will recognize this as the value of π, rounded off to two decimal places. The area of a 2″ diameter circle is 78.5% the area of a 2″ square.

Figure 2-2 Although diameter d of the circle is equal to side s of the square, its area is less

---

Force is the use of power which moves, or tries to move, any object or substance such as a rock, a door, a piston, or a quantity of fluid.

FORCE

Force is usually measured in pounds, or in the metric system, kilograms. It is derived from our measurement of weight. Ten pounds of force will support a 10 pound load. As an industrial

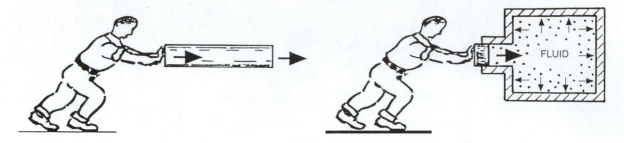

Figure 2-3 (Left) Force applied to a solid

Figure 2-4 (Right) Forced applied to a fluid

tool, hydraulics provides the means to (1) **increase** or **decrease** a force as needed, (2) **control** the amount exerted at a specific location, and (3) carry it to **several locations** where it will be used. Solid material such as steel or wood can only move the application of force to a new location; a fluid can distribute it, change its direction, and cause it to be applied in the same or different amounts at several locations.

PRESSURE

Pressure is the distribution of force over a specific, defined unit of area. In hydraulics, pressure is commonly expressed in pounds (of force) per square inch (of area), abbreviated "psi."

Given the force applied and the area over which it is distributed, we calculate the pressure by dividing the pounds of force by the number of square inches in that area. This method is illustrated in Figure 2-5. For this we use the formula

$$P = F/A \ (F \text{ divided by } A)$$

where

P = Pressure
F = Force
A = Area.

Example: What pressure results from the application of 200 lbs. of force to an area of 25 sq. in.?

$$P = F/A$$
$$P = 200/25$$
$$P = 8 \text{ psi (Ans.)}$$

Usually we are given the shape of the surface and its dimensions and have to **calculate** its area. Hydraulic cylinders and pistons are round, so we use the formula

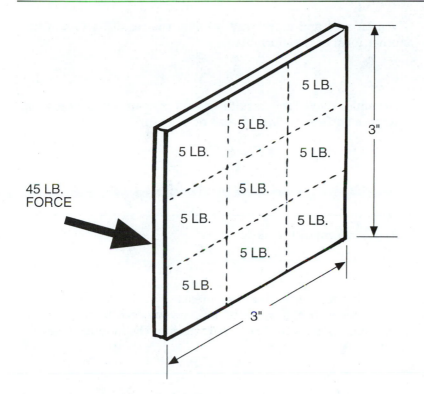

Figure 2-5 Pressure results when a force is distributed over an area

$$A = .785d^2.$$

Example: What pressure results when an 800 lb. force is applied to the face of a 4″ diameter piston? First calculate the area of the piston face:

$$A = .785d^2$$
$$A = .785(4)^2$$
$$A = .785(16) = 12.56 \text{ sq. in.}$$

Now use this figure in the formula for pressure:

$$P = F/A$$
$$P = 800/12.56$$
$$P = 63.69 \text{ psi (Ans.).}$$

We have seen that given a force and an area, we can find the resulting pressure using the formula

SOLVING FOR FORCE

$$P = F/A.$$

Given an area and a pressure, we find the resulting force using another form of that formula:

$$F = PA.$$

Example: What force results when a pressure of 250 psi is applied to a surface having an area of 4 sq. in.?

$$F = PA$$
$$F = 250(4)$$
$$F = 1000 \text{ lbs. (Ans.).}$$

SOLVING FOR AREA

Another form of the basic

$$P = F/A$$

formula is used when given the pressure available and the force required for a specific application. Now we need to calculate the area, usually of a piston, that will do the job. The formula for area is:

$$A = F/P.$$

Example: Given an available hydraulic pressure of 150 psi, what total area of piston surface will be required to develop a force of 450 lbs.?

$$A = F/P$$
$$A = 450/150$$
$$A = 3 \text{ sq. in. (Ans.).}$$

PASCAL'S LAW

The operation of all closed hydraulic systems using fluid under pressure is based upon a principle of physics known as **Pascal's Law**. This was first published in the 1600s by a brilliant Jesuit priest named Blaise Pascal, whose accomplishments also included the development of the syllable-marking system used in dictionaries, and the production of notable work in mathematical logic. The exact wording as originally stated appears to have been lost over 300 years of translations, but the main points have been preserved

Pascal's Law states that pressure on a confined fluid is transmitted equally and perpendicular to the entire surface of its container (see Figure 2-6).

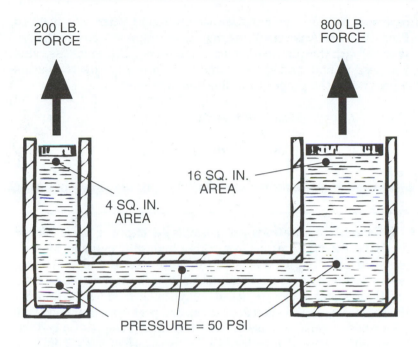

200 LB.
FORCE

800 LB.
FORCE

16 SQ. IN.
AREA

4 SQ. IN.
AREA

PRESSURE = 50 PSI

Figure 2-6 Pressure in a confined fluid acts equally on the entire surface of its container

It is important that we think of a "container" not just as a single bottle, tank, or cylinder, but as the entire surface in contact with the fluid including connecting lines or piping. With pressure transmitted equally, the total **force** developed on any surface is **proportional** to its **area**.

Two points to remember regarding Pascal's Law:

1. It holds true only if the fluid is **not flowing**. This is called a **hydrostatic system**. When fluid moves, friction causes a drop in pressure so that it becomes less downstream;
2. It does not consider the **weight of the fluid**. This is important in civil engineering with problems involving dams and water tanks, but in industrial applications, it is too small to be a factor.

Now that we know the basic formulas, we can solve several kinds of problems involving hydraulic cylinder applications. We would need to allow for the pressure drop resulting from friction before applying our calculations to actual circuitry, but for now we will assume a "perfect" system in which there is no friction.

A convenient way to remember the formulas for solving problems involving force, pressure, and area is to think of a "pyramid"

SOLVING AREA, FORCE, AND PRESSURE PROBLEMS

"FIND PROBLEM ANSWERS"

Figure 2-7 Memory aid for solving force, pressure, and area problems

representing these values. Use this pyramid when you need to **Find Problem Answers.** You may use Figure 2-7 to familiarize yourself with this method. To use this memory aid, cover the value you need to find, and the pyramid will show you how to use the other two in the proper formula. For example,

Cover F and you see PA;

Cover P and you see $\dfrac{F}{A}$;

Cover A and you see $\dfrac{F}{P}$.

CYLINDER CONSTRUCTION

Although modifications are possible to improve performance for particular applications, a working hydraulic cylinder like the one shown in Figure 2-8 is basically quite simple. Pipe or tubing is attached to the ports, and all the air in both the connecting lines and the cylinder itself is replaced by hydraulic fluid (oil). To **extend** the piston rod, the oil entering the **cap end port** is pressurized. This oil pushes against the face of the piston, and the piston rod extends with a force determined by the pressure of the oil and the area of the piston face.

$$F = PA$$

To **retract** the piston rod (pull it back), oil under pressure is pumped into the **rod end port.** Now the oil has less area to push against,

Figure 2-8 Typical cylinder construction

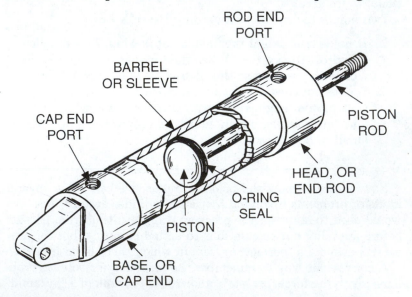

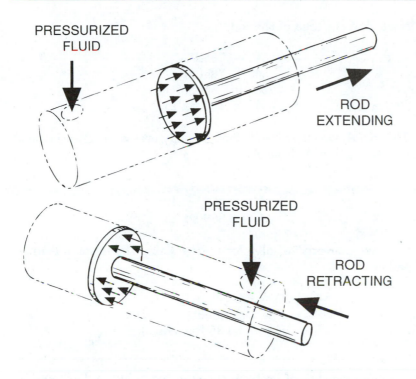

PRESSURIZED
FLUID

ROD
EXTENDING

PRESSURIZED
FLUID

ROD
RETRACTING

Figure 2-9 During retraction, the pressurized fluid (oil) has less area to push against, therefore less force is developed

because the rod itself takes up part of the area of the piston. For any given pressure then, the force exerted by the piston rod will be less when it is retracted than when it is extended. This is demonstrated in Figure 2-9.

To calculate the force developed when the rod is retracted using

$$F = PA$$

the figure used for A must be the area of the piston **minus** the area taken up by the piston rod.

Example: What force is exerted when retracting a 3″ diameter piston having a ½″ diameter rod, if the pressure of the hydraulic fluid is 150 psi?

First find the areas of the piston and rod. It is helpful to call the area of the piston A_p and the area of the rod A_r. The p and the r are called "subscripts" and are often used when we need to label several unknown quantities. To find the area of the piston:

$$A_p = .785d^2$$
$$A_p = .785(3)^2$$
$$A_p = .785(9)$$
$$A_p = 7.065 \text{ sq. in.}$$

Next, find the area of the rod:

$$A_r = .785d^2$$
$$A_r = .785(.5)^2$$
$$A_r = .785(.25)$$
$$A_r = .196 \text{ sq. in.}$$

The area A, which hydraulic fluid actually presses against, is found by subtracting A_r from A_p:

$$A = A_p - A_r$$
$$A = 7.065 - .196$$
$$A = 6.869 \text{ sq. in.}$$

Now we are ready to solve for force F using the formula learned earlier

$$F = PA$$
$$F = 150(6.869)$$
$$F = 1030.35 \text{ lbs. (Ans.).}$$

CALCULATION OF CYLINDER SIZE

A common type of problem encountered in hydraulics requires that we determine the size of cylinder needed for a specific application, given the force needed and the pressure available. For this you need to learn another mathematical operation—finding the **square root** of a number. You have already learned that when a number is multiplied by itself, it is said to be "squared," and that this operation is indicated by placing a "2" above and to the right of the number.

Finding the "square root" of a number is the reverse of "squaring" a number.

Just as we indicate a squared number with a "2" above and to the right, we indicate the square root by placing a square root sign $\sqrt{}$ over it.

$$3^2 = 9 \qquad \sqrt{9} = 3$$
$$5^2 = 25 \qquad \sqrt{25} = 5$$

It is possible to find a square root manually (that is, with pencil and paper), but it is more convenient to use a calculator or computer.

You already know the formula for finding area A of a circle when given the diameter d:

$$A = .785d^2.$$

To calculate cylinder size, use the reverse of this formula and find diameter d when given area A:

$$d = \sqrt{\frac{A}{.785}}.$$

This is done in two steps:

1. Divide the number given for area A by .785;
2. Use a calculator to find the square root.

Example: Given a pressure P of 80 psi, what size hydraulic cylinder will develop a force of 1000 lbs.? First find area A of the cylinder:

$$A = F/P$$
$$A = 1000/80$$
$$A = 12.5 \text{ sq. in.}$$

Now use this to find cylinder diameter d.

$$d = \sqrt{\frac{A}{.785}}$$

$$d = \sqrt{\frac{12.5}{.785}}.$$

$$d = \sqrt{15.924}$$

$$d = 3.99'' \text{ (Ans.)}.$$

We would probably choose a 4″ diameter cylinder for this application, or perhaps the next size larger to allow for pressure loss due to friction.

MECHANICAL LEVERAGE

Even without actually studying mechanics, we have all made use of the principle of leverage. We know we can exert many pounds of force at the end of a crowbar. In using a claw hammer to pull nails, we realize we can pull harder if we hold the end of the handle rather than the middle. As Figure 2-10 illustrates, a heavy weight on a board can be balanced by a lighter one placed farther away from the support.

Leverage results from a force applied at some distance from a support, called a **fulcrum**. There are three basic types of "levers," classified by the relative locations of the **applied force** (F_1), the **resulting**, or **resultant force** (F_2), and the **fulcrum**. The effectiveness of each force depends on both the **amount** of force and its **distance from the fulcrum**.

Figure 2-10 "Leverage" in a mechanical system permits a small force to overcome a larger one

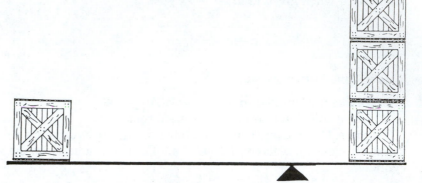

Each lever shown in Figure 2-11 is acted upon by two forces, F_1 and F_2, placed at distances D_1 and D_2 from the fulcrum. In each example, F_1 tends to rotate the lever about the fulcrum in a **clockwise** direction; F_2 tends to rotate it **counter-clockwise**.

Consider **LEVER A.** You would agree that if forces F_1 and F_2 are equal and distances D_1 and D_2 are equal, the lever is "in balance" and will not rotate in either direction. Furthermore, you probably know from experience that if F_1 is less than F_2, it must be farther from the fulcrum for the lever to be in balance.

The tendency of a force to rotate a lever about a fulcrum is called a "moment" and is calculated by multiplying the amount of force by its distance from the fulcrum. For the lever to be in balance, the clockwise moments and the counter-clockwise moments must be equal.

$$(F_2)(D_2) = (F_1)(D_1)$$

Given F_1, D_1, and D_2, we solve force F_2 using the formula

$$F_2 = \frac{F_1 D_1}{D_2} .$$

Example: Find the force F_2 needed to balance **LEVER A** in Figure 2-11 if F_1 is 8 lbs., D_1 is 20 in., and D_2 is 16 in.

$$F_2 = \frac{F_1 D_1}{D_2}$$

$$F_2 = \frac{(8)(20)}{16}$$

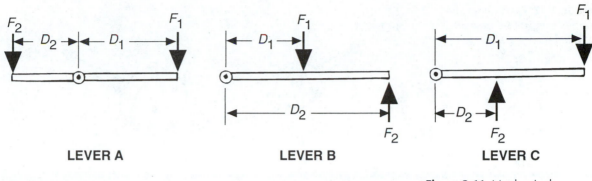

LEVER A   LEVER B   LEVER C

Figure 2-11 Mechanical leverage

$$F_2 = \frac{160}{16}$$

$$F_2 = 10 \text{ lbs. (Ans.).}$$

Given F_1, F_2, and D_1 we can find D_2 using the formula:

$$D_2 = \frac{F_1 D_1}{F_2}.$$

Example: Find distance D_2 needed to balance the **LEVER A** in Figure 2-11 if F_1 is 6 lbs., D_1 is 15 in., and F_2 is 9 lbs.

$$D_2 = \frac{F_1 D_1}{F_2}$$

$$D_2 = \frac{(6)(15)}{(9)}$$

$$D_2 = \frac{(90)}{(9)}$$

$$D_2 = 10 \text{ inches (Ans.).}$$

These same formulas would be used to find the forces and distances for **LEVER B** and **LEVER C** in Figure 2-11.

HYDRAULIC LEVERAGE

We have just learned how a force can be increased by means of mechanical leverage. We shall now see how this can be accomplished using a **confined fluid,** applying what we know about area, force, and pressure. A force applied to a surface develops a **pressure,** which depends on the amount of the force and the size of the

Figure 2-12 Fluid under pressure is used to increase an applied force

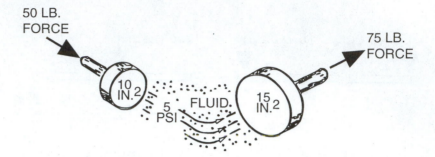

surface. If the surface is now made to push against a **confined fluid**, the fluid is said to be **pressurized**. The pressure of the **fluid** will be the same as that of the **surface**.

If the pressurized **fluid** is then made to push against a second **surface**, a second **force** will be developed which depends upon the amount of **pressure** and the **size** of the second surface. The amount of increase or decrease depends upon the relative **areas** of the two surfaces. In Figure 2-12, the **input** force of 50 lbs. on a surface of 10 square inches develops a pressure of 5 psi. This pressure, by means of a **confined fluid**, acts upon a surface of 15 square inches to produce an **output** force of 75 lbs. The ratio of the two forces is exactly the same as the ratio of the two areas, and we can express this in the formula:

$F = PA$

$$\frac{F_2}{F_1} = \frac{A_2}{A_1}.$$

This means, "Output force F_2 is to input force F_1 as the area of the output piston surface A_2 is to the input piston surface A_1." If A_2 is twice as large as A_1 then F_2 will be twice as much as F_1. If A_2 is three times as large as A_1 then F_2 will be three times as much as F_1. The **ratio** is the same for the forces as it is for the areas. We use another form of this formula to find force F_2 if given F_1, A_1, and A_2.

$$F_2 = \frac{F_1 A_2}{A_1}$$

This multiplication of force will usually, but not always (as we shall see in Chapter 9), be accomplished using two hydraulic cylinders.

Example: What will be the output force at F_2 if input force F_1 is 20 lbs., piston area A_1 is 2 sq. in., and piston area A_2 is 16 sq. in.? (see Figure 2-13.)

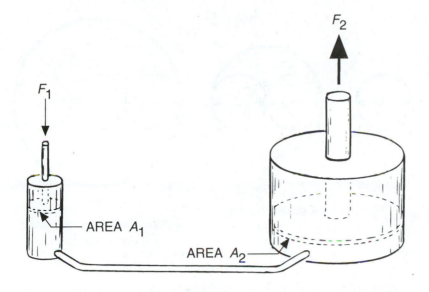

Figure 2-13 The ratio between the forces exerted on the piston rods is the same as the ratio between the areas of the piston faces

$$F_2 = \frac{F_1 A_2}{A_1}$$

$$F_2 = \frac{(20)(16)}{(2)}$$

$$F_2 = \frac{320}{2}$$

$$F_2 = 160 \text{ lbs. (Ans.).}$$

Since the ratio of the two forces is always the same as the ratio of the areas, we have a much simpler way to find F_2. Notice that the area A_2 is 8 times the size of area A_1. Therefore, force F_2 has to be 8 times F_1, which was given as 20 lbs. Multiplying, we find our answer is again 160 lbs.

A common practical problem in hydraulics requires that we select two cylinders that will increase a given force to some amount. We are given input force F_1 and the needed output force F_2. We divide F_2 by F_1 to find their **ratio**, which is the **multiplication factor** of the output as compared to the input. Put another way, we divide the input force into the output force to determine how many times it has to be increased to get the output force we need.

$$\frac{F_2}{F_1} = \text{Multiplication Factor}$$

Figure 2-14 The area of a circle increases as the square of the change in diameter

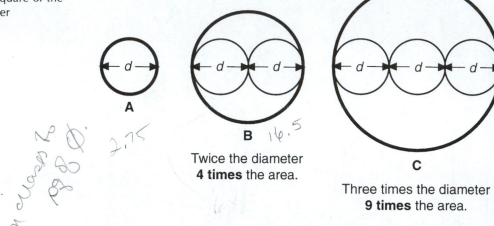

A

B

Twice the diameter
4 times the area.

C

Three times the diameter
9 times the area.

We know that the ratio of the **forces** of the two cylinders is the **same** as the ratio of the **areas**. The size of a hydraulic cylinder, however, is given as its **diameter**.

The ratio between the areas of two cylinders is the same as the ratio of the squares of their diameters.

To illustrate: Consider two cylinders, C_2 and C_1, having diameters of 9″ and 3″ respectively. The ratio of the diameters, C_2 to C_1 is obviously **3**. Using the formula for area,

$$A = .785d^2,$$

we find that the area of C_2 is 63.585 sq. in. and the area of C_1 is 7.065 sq. in. The area of C_2 is **9 times** that of C_1, not 3 times. This is 3^2, the **ratio between the squares of their diameters**.

This permits us to simplify our calculations when solving for output force in a given system, or for cylinder sizes needed for a particular application. We use **ratios** rather than actual diameters or areas, and we find the ratio between two numbers by **dividing**.

Example: If diameter d_1 of cylinder C_1 is 2.5″, and diameter d_2 of cylinder C_2 is 4.0″, what is the ratio between their areas? (see Figure 2-15.)

$$\frac{A_2}{A_1} = \frac{(d_2)^2}{(d_1)^2}$$

$$\frac{A_2}{A_1} = \frac{(4.0)^2}{(2.5)^2}$$

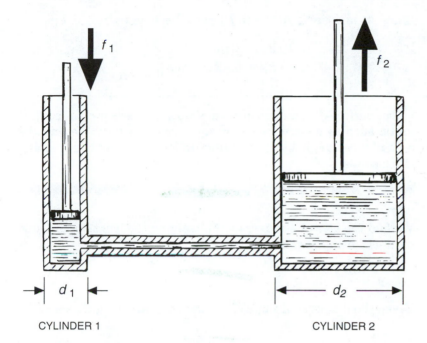

CYLINDER 1   CYLINDER 2

$$\frac{A_2}{A_1} = \frac{16}{6.25}$$

$$\frac{A_2}{A_1} = 2.56 \text{ (Ans.)}$$

Knowing that area A_2 is 2.56 times the size of A_1, we can find any output force F_2 just by multiplying F_1 by this figure. Remember: The ratio of the **forces** is the same as the ratio of the **areas**.

Example: If input force F_1 is 15 lbs., the diameter of **Cylinder C$_1$** is 2.5″, and the diameter of **Cylinder C$_2$** is 5″, what force is exerted at F_2? (see Figure 2-15.) First find the ratio between the areas.

$$\frac{A_2}{A_1} = \frac{(d_2)^2}{(d_1)^2}$$

$$\frac{A_2}{A_1} = \frac{(5)^2}{(2.5)^2}$$

$$\frac{A_2}{A_1} = \frac{25}{6.25}$$

$$\frac{A_2}{A_1} = 4$$

Since A_2 is 4 times A_1, force F_2 must be 4 times F_1.

$$F_2 = 4(15)$$
$$F_2 = 60 \text{ lbs. (Ans.).}$$

CHAPTER SUMMARY

Flat surfaces, regardless of their shape, are measured in "square" units, such as square inches or square millimeters. We find the area of a rectangle by multiplying its length by its width, using the formula:

$$A = LW.$$

We find the area of a square by multiplying the length of one side by itself, using the formula:

$$A = s^2.$$

Given the diameter of a circle, we find the area using the formula:

$$A = .785d^2.$$

Force is the use of power which moves, or tries to move, any object or substance. Pressure is the distribution of force over an area, and is calculated using the formula:

$$P = F/A.$$

Using the **Find Problem Answers** pyramid as a memory aid we solve problems involving **force**, **pressure**, and **area**, with the formulas:

$$F = PA$$
$$P = F/A$$
$$A = F/P.$$

Pascal's Law states that **pressure** on a confined fluid is transmitted **equally** and **perpendicular** to the entire surface of its container. For a given pressure, the force exerted on a piston rod during **retraction** is **less** than during **extension** because the hydraulic fluid (oil) has less area to push against.

Given the area of a circle, we find its diameter using the formula:

$$d = \sqrt{\frac{A}{.785}}.$$

The tendency of a force to rotate a lever about a fulcrum is called a **moment** and is calculated by multiplying the **amount** of force by its **distance** from the **fulcrum**. Forces are in balance when the **clockwise moments** and **counter-clockwise moments** are equal.

Hydraulic leverage results when a **pressurized, confined fluid** acts upon piston surfaces. The forces thus produced are proportional to the **areas** of the surfaces or to the **squares** of the piston **diameters**.

PROBLEMS

2.1 Find the area of a rectangle .78″ wide by 2.3″ long.

2.2 Find the area of a rectangle 2.8″ wide by 3.75″ long.

2.3 Find the area of a rectangle 4.25″ long by 1.05″ wide.

2.4 Find the area of a rectangle 6.25″ long by 3.20″ wide.

2.5 Find the area of a square which measures 4.6″ per side.

2.6 Find the area of a square which measures .75″ per side.

2.7 Find the area of a square which measures 8.50″ per side.

2.8 Find the area of a square which measures 4.75″ per side.

2.9 Find the area of a 4.30″ diameter circle.

2.10 Find the area of a 5.18″ diameter circle.

2.11 Find the area of a .92″ diameter circle.

2.12 Find the area of an 8.0″ diameter circle.

2.13 What pressure results when 125 lbs. of force is applied to an area of 5 sq. in.?

2.14 What pressure results when 260 lbs. of force is applied to an area of 1.30 sq. in.?

2.15 What pressure results when 300 lbs. of force is applied to a 3″ by 50″ rectangular surface?

2.16 What pressure results when 210 lbs. of force is applied to a square surface measuring .50″ per side?

2.17 What pressure results when 500 lbs. of force is applied to a square surface measuring .80″ per side?

2.18 What pressure results when 150 lbs. of force is applied to a 3.00″ diameter circular surface?

2.19 What pressure results when 750 lbs. of force is applied to a .75″ diameter circular surface?

2.20 What pressure results when 720 lbs. of force is applied to the rod of a 1.75″ diameter piston?

2.21 What pressure results when 850 lbs. of force is applied to the rod of a 2.25″ diameter piston?

2.22 What force results when 200 psi is applied to a surface having an area of 2.50 sq. in.?

2.23 What force results when 80 psi is applied to a surface having an area of 60 sq. in.?

2.24 What force results when 120 psi is applied to the face of a 2.25″ diameter piston?

2.25 What force results when 400 psi is applied to the face of a 1.75″ diameter piston?

2.26 With what force will a piston rod extend when oil at 200 psi is directed into the cap end port of a 3″ diameter hydraulic cylinder?

2.27 With what force will a piston rod extend when oil at 500 psi is directed into the cap end port of a 7.50″ diameter hydraulic cylinder?

2.28 Given a hydraulic system with an oil pressure of 250 psi, what area of piston surface will be required to develop a force of 1000 lbs. on the rod?

2.29 Given a hydraulic system with an oil pressure of 250 psi, what area of piston surface will be required to develop a force of 2400 lbs. on the rod?

2.30 Given a hydraulic system with an oil pressure of 750 psi, what area of piston surface will be required to develop a force of 22,500 lbs. on the rod?

2.31 You have a 4″ hydraulic cylinder with a .75″ diameter piston rod in a system operating at 100 psi.
(a) With what force will the rod extend when oil is directed into the cap end port?
(b) With what force will the rod retract when oil is directed into the rod end port?

2.32 You have a 6.50″ hydraulic cylinder with a 1.00″ diameter piston rod in a system operating at 250 psi.
(a) With what force will the rod extend when oil is directed into the cap end port?
(b) With what force will the rod retract when oil is directed into the rod end port?

2.33 You have a 7.75″ hydraulic cylinder with a 1.50″ diameter piston rod in a system operating at 330 psi.
(a) With what force will the rod extend when oil is directed into the cap end port?
(b) With what force will the rod retract when oil is directed into the rod end port?

2.34 You have a 1.25″ hydraulic cylinder with a .25″ diameter piston rod in a system operating at 80 psi.
(a) With what force will the rod extend when oil is directed into the cap end port?
(b) With what force will the rod retract when oil is directed into the rod end port?

2.35 Given a hydraulic system with an oil pressure of 85 psi, what size cylinder will you need to support a 3800 lb. automobile resting on the end of its rod?

2.36 Given a hydraulic system with an oil pressure of 110 psi, what size cylinder will you need to develop a force of 450 lbs. on the end of its rod?

2.37 Given a hydraulic system with an oil pressure of 230 psi, what size cylinder will you need to develop a force of 1000 lbs. on the end of its rod?

2.38 Given a hydraulic system with an oil pressure of 150 psi, what size cylinder will you need to develop a force of 875 lbs. on the end of its rod?

Refer to Figure 2-16 for solving problems 2.39 through 2.50.

2.39 What force F_2 results if $F_1 = 35$ lbs., piston area $A_1 = 7$ sq. in., and piston area $A_2 = 24.5$ sq. in.?

2.40 What force F_2 results if $F_1 = 150$ lbs., piston area $A_1 = 1.25$ sq. in., and piston area $A_2 = 18$ sq. in.?

2.41 What force F_2 results if $F_1 = 13$ lbs., piston area $A_1 = 6.5$ sq. in., and piston area $A_2 = 24$ sq. in.?

2.42 What force F_2 results if $F_1 = 210$ lbs., piston area $A_1 = 3.5$ sq. in., and piston area $A_2 = 70$ sq. in.?

2.43 If cylinder diameter $d_1 = 1.5''$ and cylinder diameter $d_2 = 15''$, what is the ratio between their piston areas?

2.44 If cylinder diameter $d_1 = 3''$ and cylinder diameter $d_2 = 12''$, what is the ratio between their piston areas?

2.45 If cylinder diameter $d_1 = 4''$ and cylinder diameter $d_2 = 24''$, what is the ratio between their piston areas?

2.46 If cylinder diameter $d_1 = .75''$ and cylinder diameter $d_2 = 2.25''$, what is the ratio between their piston areas?

2.47 What force F_2 results when $F_1 = 20$ lbs., diameter $d_1 = 2''$, and diameter $d_2 = 8'''$?

2.48 What force F_2 results when $F_1 = 80$ lbs., diameter $d_1 = 1.0''$, and diameter $d_2 = 7.0'''$?

2.49 What force F_2 results when $F_1 = 55$ lbs., diameter $d_1 = 3.5''$, and diameter $d_2 = 12'''$?

2.50 What force F_2 results when $F_1 = 17$ lbs., diameter $d_1 = 1.6''$, and diameter $d_2 = 9.0'''$?

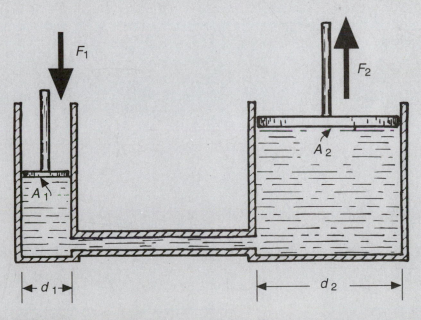

Figure 2-16

Volume, Capacity, and Fluid Flow

3

In Chapter 2, we learned how fluid in a confined system is used to produce force or pressure. By selecting appropriate sizes of hydraulic cylinders, we are able to develop **hydraulic leverage** and convert a given force to the level needed to perform some needed function. There was no mention of how **fast** a piston rod would extend when acted upon by pressurized oil—those formulas told us only how much **force** was developed. This brings us to a very important point to remember in the study of hydraulics:

The rate of piston rod extension or retraction is independent of pressure.

Extension rate is normally expressed in inches per minute, which we write as **ipm** or **in./min.** To determine how **fast** the rod extends we need to know the **flow rate** at which hydraulic oil is directed into the cylinder port. This is expressed in **gallons per minute (GPM)** and is usually, but not always, determined by the capability of the pump. As the incoming oil fills the cylinder between the piston and either end, the rod is forced to move to make room. We determine the **rate** at which it extends or retracts using calculations based upon **volume, capacity,** and **fluid flow.**

pressurized air 20 psi

VOLUME

Volume is the measure of an enclosed space or solid figure, and regardless of the shape, is given in "cubic" units such as cubic inches or cubic millimeters.

There are several formulas used for finding a volume depending upon its shape, whether it be a cube, rectangular solid, pyramid, sphere, prism, cylinder, or a combination. For some shapes, those having a **constant cross-sectional area**, the calculation is quite simple. We find the area, then multiply this figure by the height:

$$V = AH$$

where

$V = $ Volume in cubic inches
$A = $ Area of the cross-section
$H = $ Height.

Rectangular solids and cylinders are two figures having constant cross-sectional areas. This means that in Figure 3-1, any "slice" taken perpendicular to the height has the same size and shape. We could calculate the volume of a rectangular solid by first finding the area of the base and then multiplying by the height, but it is more convenient to combine the two steps into a single formula:

$$V = (L)(W)(H), \text{ or } LWH$$

where

$V = $ Volume
$L = $ Length of the base
$W = $ Width of the base
$H = $ Height.

Example: What is the volume of a 20″ high rectangular solid whose base measures 7″ × 4″?

$$V = LWH$$
$$V = (7)(4)(20)$$
$$V = 560 \text{ cu. in. (Ans.)}.$$

We could calculate the volume of a cylinder by first finding the area of the circular cross-section and then multiplying it by the height. Again, however, it is more convenient to combine the steps in a single formula:

$$V = .785d^2h$$

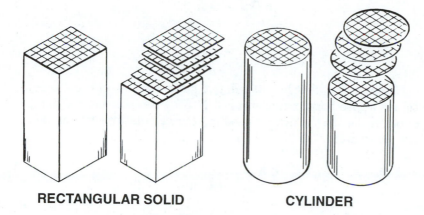

RECTANGULAR SOLID   **CYLINDER**

Figure 3-1 Solid figures having constant cross-sectional areas. Each "slice" has the same size and shape

where

$$V = \text{Volume}$$
$$d = \text{cylinder diameter}$$
$$h = \text{cylinder height.}$$

Example: What is the volume of a 6″ high cylinder whose diameter is 4″?

$$V = .785d^2h$$
$$V = .785(4)^2(6)$$
$$V = .785(16)(6)$$
$$V = 75.36 \text{ cu. in. (Ans.).}$$

CAPACITY

Since liquids have no specific length, width, or height dimensions to measure, we normally express their quantity in terms of standard container sizes such as ounces, quarts, liters, or gallons. A container is said to have a **capacity** of one gallon, or one ounce, or one liter, and it need have no specific shape. Each unit of capacity, however, corresponds to a volume which can be expressed in **cubic units.** We need to know just one equivalent value, and it is important enough that you should commit it to memory:

One gallon = 231 cubic inches.

When a given amount of liquid is poured into a container having a constant cross-sectional area, it is a simple matter to calculate to what height the container will be filled. Here we use another form of the formula

$$V = AH:$$

$$H = \frac{V}{A}$$

Example: To what height will two gallons fill a rectangular container whose bottom measures 6″ × 6″? First convert gallons to cubic inches, so the figures used in your calculations will all be in the **same units**.

$$V = 2(231)$$
$$V = 462 \text{ cu. in.}$$

Next find the **cross-sectional area** of the base.

$$A = s^2$$
$$A = 6(6)$$
$$A = 36 \text{ sq. in.}$$

Finally, divide the volume of the liquid by the area of the base to find the height.

$$H = \frac{V}{A}$$

$$H = \frac{462}{36}$$

$$H = 12.84 \text{ in. (Ans.).}$$

Note that this is twice the height of the cube shown in Figure 3-2, which has a volume equivalent to only one gallon.

Figure 3-2 One gallon has a volume equal to 231 cubic inches

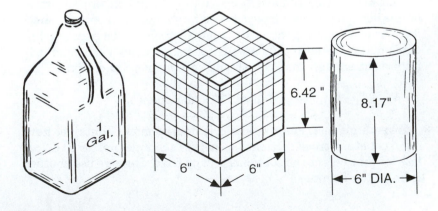

After first converting gallons to cubic inches (multiplying by 231) we can use a variation of the formula

$$V = LWH,$$

which we have already learned, to find the height.

$$H = \frac{V}{LW}$$

Example: To what height will 3.5 gallons fill a container whose bottom measures 8″ × 7″?

$$V = 3.5(231)$$
$$V = 808.5 \text{ cu. in.}$$

$$H = \frac{V}{LW}$$

$$H = \frac{808.5}{8(7)}$$

$$H = \frac{808.5}{56}$$

$$H = 14.44 \text{ in. (Ans.).}$$

If the container has a cylindrical shape, we use a similar approach. Now, however, we begin with the formula for the volume of a cylinder but use it in another form to find the height:

$$V = .785d^2h$$

is changed to

$$h = \frac{V}{.785d^2}$$

Example: To what height will 2 gallons fill a 6″ diameter cylindrical container? First convert gallons to cubic inches so the figures used in your calculations will all be in the **same units**:

$$V = 2(231)$$
$$V = 462 \text{ cu. in.}$$

Figure 3-3 Each cylinder has a volume of 231 cubic inches, which is the same as the capacity of a one-gallon jug

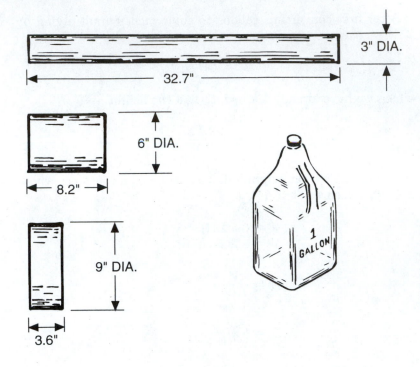

$$h = \frac{V}{.785d^2}$$

$$h = \frac{462}{.785(6)^2}$$

$$h = \frac{462}{.785(36)}$$

$$h = \frac{462}{28.26}$$

$$h = 16.34'' \text{ (Ans.).}$$

Note that this is twice the height of the cylinder shown in Figure 3-2, which has a volume equivalent to only one gallon. As shown in Figure 3-3, a 231 cubic-inch cylinder can have many combinations of height and diameter dimensions.

PISTON ROD EXTENSION　　　When oil is pumped into the cap end port of a hydraulic cylinder, it pushes against the face of the piston to extend the rod. In Figure 3-4, you can see that the addition of oil will move the piston

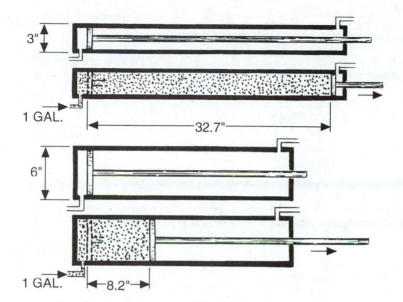

some distance depending upon (1) the amount of oil added and (2) the size of the cylinder. One gallon pumped into a 3″ cylinder occupies a space whose height, or length, is 32.7″. The rod must therefore extend 32.7″. One gallon pumped into a 6″ cylinder would extend the rod only 8.2″. This assumes, of course, that the pressure developed is enough to overcome any opposition to rod movement.

If oil is pumped into the **rod** end port, where part of the volume of the cylinder is taken up by the rod itself, we must take this into account in our calculations. We still use the formula which says that

PISTON ROD RETRACTION

$$\text{volume} = \text{area} \times \text{height},$$

but now the figure used for A is the area of the inside of the cylinder **minus** the area of the rod. We find A using

$$A = .785(d_{cyl})^2 - .785(d_{rod})^2$$

where

$$A = \text{Area oil pushes against}$$
$$d_{cyl} = \text{cylinder diameter}$$
$$d_{rod} = \text{rod diameter.}$$

Example: You have a 3″ hydraulic cylinder with a 1.25″ diameter piston rod. How far will the rod retract when .5 gallon of oil is

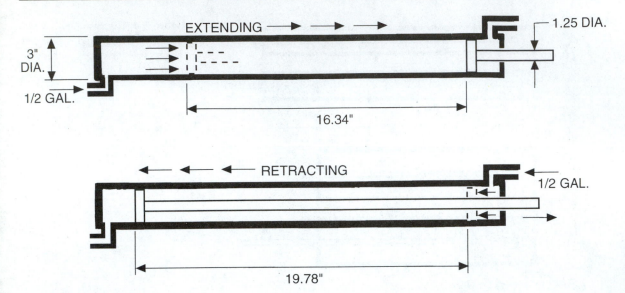

Figure 3-5 A piston moves farther when retracting than when extending, for the same amount of fluid input

pumped into its rod end port? First convert the .5 gallon of oil to cubic inches so the figures used in your calculations will all be in the **same units.**

$$V = .5(231)$$
$$V = 115.5 \text{ cu. in.}$$

Next, solve for A.

$$A = .785(d_{cyl})^2 - .785(d_{rod})^2$$
$$A = .785(3)^2 - .785(1.25)^2$$
$$A = .785(9) - .785(1.5625)$$
$$A = 7.065 - 1.226$$
$$A = 5.839 \text{ sq. in.}$$

Figure 3-6 When leverage is used to increase force, the smaller force must travel a greater distance

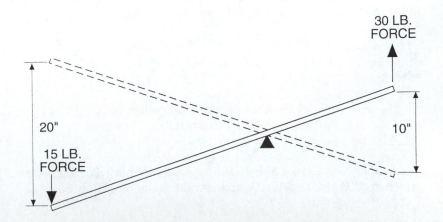

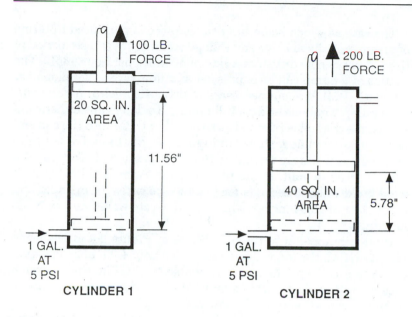

Figure 3-7 Doubling the piston area doubles the force exerted, but reduces the length of rod travel by half

Now use these figures in the formula for height.

$$h = \frac{V}{A}$$

$$h = \frac{115.5}{5.839}$$

$$h = 19.78'' \text{ (Ans.).}$$

As shown in Figure 3-5, ½ gallon of oil pumped into the **cap end** port of a 3″ cylinder extends the piston rod only 16.34″.

In Chapter 2, we compared hydraulic cylinders with mechanical levers, and learned how a force can be increased by hydraulic or mechanical means. In a mechanical system it depends upon the lengths of **lever arms**; in a hydraulic system it varies according to the **piston areas**.

Now we have another point of similarity. When we use a lever to increase a force, the smaller force must travel a **greater distance**. Compare this with the cylinders shown in Figure 3-4, keeping in mind that the multiplication of force is proportional to the **areas**, not the diameters, of the piston faces. The cross-sectional area of a 6″ cylinder (28.26 sq. in.) is **four times** that of a 3″ cylinder (7.065 sq. in.). The input force is multiplied four times, but for an input of one gallon of oil, the piston moves only one-fourth as far. Comparing **areas** rather than diameters in Figure 3-7, we see that doubling the area doubles the output force but reduces the distance traveled by half.

EXTENSION RATE

The **rate** at which something takes place is measured in terms of an **amount, distance,** or any **change** occurring over a set period of **time**. A car can be driven at a **rate** of 40 miles per hour **(mph)**. The output of a pump can be expressed as a **flow rate** of 2 gallons per minute **(GPM)**. It is not necessary to drive 40 miles or for an hour, or to pump 2 gallons or for a full minute, to determine either rate.

Assume that oil is pumped at a rate of ½ GPM into the cap end port of the 3″ cylinder shown in Figure 3-4. At the end of one minute, the rod will have extended 16.34″. It extends, therefore, at a rate of **16.34 in./min.** If the pumping rate were 1 GPM, then it would take only 30 seconds for ½ gallon to enter the cylinder and move the rod 16.34 inches. The extension rate would now be **32.68 in./min.**

If oil is pumped into the 6″ cylinder of Figure 3-4 at the same rate, ½ GPM, the rod will have extended only 4.09″ after one minute. Again, if the pumping rate is doubled to 1 GPM, the extension rate also doubles, to 8.18 in./min. We say the extension rate is **directly proportional** to the **flow rate**.

In each example, note that the **pressure** was never mentioned. As long as the pressure is enough to move the piston, the extension rate depends only upon the cylinder diameter and the pumping, or flow rate of the incoming oil. The formula for finding the extension rate is very similar to the one for determining how far a given amount of oil moves the piston:

$$R = \frac{231(g)}{.785d^2}$$

where

$$R = \text{Rod extension rate}$$
$$g = \text{oil flow rate (GPM)}$$
$$d = \text{cylinder diameter.}$$

Example: At what rate will a piston rod extend when oil is pumped into the cap end port of a 4″ cylinder at 1.5 GPM?

$$R = \frac{231(g)}{.785d^2}$$

$$R = \frac{231(1.5)}{.785(4)^2}$$

$$R = \frac{346.5}{.785(16)}$$

$$R = \frac{346.5 \ GPM}{12.56 \ in^2}$$

$$R = 27.59 \text{ in./min. (Ans.).}$$

RETRACTION RATE

As we have seen (Figure 3-5), a given amount of oil pumped into the **rod end port** will **retract** a piston rod farther than the same amount pumped into the **cap end port** will **extend** it. Using the figures given in the illustration, a flow rate of ½ GPM would retract the rod at a rate of 19.78 in./min. but would extend it at only 16.34 in./min. The formula used for calculations must account for the volume taken up by the piston rod, just as was done when calculating retraction **distance**. The volume to be filled with oil is found by subtracting the area of the rod from the area of the cylinder, then by multiplying what remains by the height. A hollow cylinder, just like a rectangular solid or a solid cylinder, is a figure with a constant cross-sectional area. The "hole" in Figure 3-8 represents the space taken up by the piston rod.

This is another type of problem that could be solved using a single formula, but is easier to follow if we do it in two steps. First find the area of the piston surface that remains after the rod area is subtracted, using the formula

$$A = .785(d_{cyl})^2 - .785(d_{rod})^2.$$

Then use A in the formula for rod retraction rate:

$$R = \frac{231(g)}{A}$$

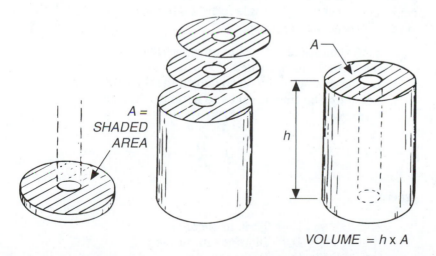

$A =$
SHADED
AREA

A

h

VOLUME = $h \times A$

Figure 3-8 Subtracting the volume taken by a piston rod leaves a solid figure with a constant cross-sectional area

where

$$R = \text{Rod retraction rate}$$
$$g = \text{oil flow rate (GPM)}$$
$$A = \text{Cyl. Area} - \text{Piston Area.}$$

Example: You have a 4″ cylinder with a 1.15″ diameter piston rod. At what rate will the rod retract if oil is pumped into the rod end port at 1 GPM?

$$A = .785(d_{cyl})^2 - .785(d_{rod})^2$$
$$A = .785(4)_2 - .785(1.5)^2$$
$$A = .785(16) - .785(2.25)$$
$$A = 12.56 - 1.77$$
$$A = 10.79 \text{ sq. in.}$$

Now use this in the formula for retraction rate.

$$R = \frac{231(g)}{A}$$

$$R = \frac{231(1)}{10.79}$$

$$R = 21.41 \text{ in./min. (Ans.).}$$

ACTUATION TIME

Another type of problem requires that we determine how many seconds it will take for a piston rod to extend or retract a specific distance. The cylinder size and the oil flow rate are given. This could be solved using a single formula, but it is easier to follow if we use a three-step approach:

1. Find the rate of extension or retraction in **in./min.**, using the formula just learned;
2. Divide this figure by 60 to determine the rate in **seconds**;
3. Solve for the time it takes to extend or retract using the formula:

$$T = \frac{D}{R_s}$$

where

$$T = \text{Time in seconds}$$
$$D = \text{Distance in inches}$$
$$R_s = \text{Extension rate, in./sec.}$$

Example: In how many seconds will the piston rod of a 4.5″ cylinder extend 7″ when oil is pumped into the cap end port at .75 GPM? First find the extension rate.

$$R = \frac{231(g)}{.785d^2}$$

$$R = \frac{231(.75)}{.785(4.5)^2}$$

$$R = \frac{173.25}{.785(20.25)}$$

$$R = \frac{173.25}{15.90}$$

$$R = 10.90 \text{ in./min.}$$

Divide by 60 to find the rate in in./sec.

$$R_s = 10.90/60$$
$$R_s = .182 \text{ in./sec.}$$

Use this figure in the formula for time:

$$T = \frac{D}{R_s}$$

$$T = \frac{7}{.182}$$

$$T = 38.46 \text{ sec. (Ans.).}$$

CHAPTER SUMMARY

The rate of piston rod extension or retraction is **independent of pressure.** Rod extension rate is normally expressed in inches per minute, which we write as **ipm** or **in./min.** Fluid flow is expressed in gallons per minute, which we write as **GPM.**

Volume is the measure of an enclosed space or solid figure, and is expressed in **cubic units.** To find the volume of any object having a **constant cross-sectional area,** multiply this area by the **height:**

$$V = AH.$$

To find the volume of a **rectangular solid,** use the formula:

$$V = LWH.$$

To find the volume of a **cylinder,** use the formula:

$$V = .785d^2h.$$

A **capacity** of **one gallon** is equal to a **volume** of **231 cubic inches.** When given a problem involving capacities and volumes, first **convert** the capacities **(gallons)** to volumes **(cubic inches).**

To find the height to which a given amount of liquid will fill any object having a **constant cross-sectional area,** use the formula:

$$H = \frac{V}{A}.$$

To find the height to which a given amount of liquid will fill a **rectangular solid** use the formula:

$$H = \frac{V}{LW}.$$

To find the height to which a given amount of liquid will fill a **cylinder** use the formula:

$$h = \frac{V}{.785d^2}.$$

To find the distance a piston rod will **extend** when a given amount of fluid is pumped into the **cap end port** of a hydraulic cylinder, use the formula:

$$h = \frac{V}{.785d^2}.$$

To find the distance a piston rod will **retract** when a given amount of fluid is pumped into the **rod end port** of a hydraulic cylinder, use a three-step approach:

1. If the amount of fluid is given in gallons, multiply by 231 to convert to cubic inches;
2. Find the effective area of the piston surface by deducting the area taken by the rod, using the formula

$$A = .785(d_{cyl})^2h - .785(d_{rod})^2h;$$

3. Use these figures in the formula for height:

$$h = \frac{V}{A}.$$

To find the rate at which a piston rod will **extend**, given the **rate** at which fluid is pumped into the **cap end port** of a hydraulic cylinder, use the formula:

$$R = \frac{231(g)}{.785d^2}.$$

To find the rate at which a piston rod will **retract**, given the **rate** at which fluid is pumped into the **rod end port** of a hydraulic cylinder, use a two-step approach:

1. Find the effective area of the piston by deducting the area taken by the piston rod, using the formula:

$$A = .785(d_{cyl})^2h - .785(d_{rod})^2h;$$

2. Use this in the formula for rate of retraction:

$$R = \frac{231(g)}{A}.$$

To find the **time** required for a piston rod to **extend** or **retract** a **specific distance**, use a three-step approach:

1. Solve for the extension or retraction **rate** in **inches per minute** (R);
2. Divide this figure by **60** to convert to **inches per second** (R_s);
3. Solve for time in **seconds** using the formula:

$$T = \frac{D}{R_s}$$

PROBLEMS

3.1 Which of the changes listed would increase the rate of piston rod extension?
   (a) Increase the piston rod diameter
   (b) Decrease the cylinder diameter ✓
   (c) Increase the pressure of the fluid
   (d) Increase the GPM input to the cap end port ✓
   (e) Pump oil into the rod end port rather than the cap end port
   (f) Decrease the piston face area ✓
   (g) Decrease the pressure of the fluid
   (h) Increase the cylinder diameter

3.2   Which of the following objects has a shape with a uniform cross-sectional area?
(a) a broom handle
(b) an ice cream cone
(c) a light bulb
(d) a section of copper tubing
(e) a wood lead pencil
(f) a soft drink bottle
(g) a hammer handle

3.3   What is the volume of a box which measures 7.5″ long by 3″ wide by 14″ high?

3.4   What is the volume of a box which measures 6″ long by 2.25″ wide by 11″ high?

3.5   What is the volume of a box which measures 3.5″ long by 1.2″ wide by 17″ high?

3.6   What is the volume of a box which measures 12″ long by 11″ wide by 15.5″ high?

3.7   What is the volume of a 3.5″ diameter by 4″ high cylinder?

3.8   What is the volume of a 4″ diameter by 3.5″ high cylinder?

3.9   What is the volume of a 2.75″ diameter by 9″ high cylinder?

3.10   What is the volume of a 1.6″ diameter by 4.8″ high cylinder?

3.11   How many cubic inches are in 2.5 gallons?

3.12   How many cubic inches are in 1.8 gallons?

3.13   How many cubic inches are in $\frac{8}{10}$ of a gallon?

3.14   How many cubic inches are in $\frac{3}{4}$ of a gallon?

3.15   To what height will 1 gallon fill a rectangular container whose bottom measures 2.25″ by 3″?

3.16   To what height will 2.2 gallons fill a rectangular container whose bottom measures 6.5″ by 8″?

3.17   To what height will $\frac{7}{10}$ gallon fill a rectangular container whose bottom measures 5″ by 5″?

3.18   To what height will 1.5 gallons fill a rectangular container whose bottom measures 4″ by 6″?

3.19   To what height will 2.1 gallons fill a 4″ diameter cylindrical container?

3.20   To what height will 3.5 gallons fill a 9″ diameter cylindrical container?

3.21   To what height will 1.65 gallons fill a 4.4″ diameter cylindrical container?

3.22   How far will the piston rod of a 3.5″ cylinder extend if 1.3 gallons of oil is pumped into the cap end port?

3.23   How far will the piston rod of a 4.5″ cylinder extend if $\frac{3}{4}$ gallon of oil is pumped into the cap end port?

3.24   How far will the piston rod of a 3.75″ cylinder extend if .5 gallon of oil is pumped into the cap end port?

3.25   How far will the piston rod of a 6″ cylinder extend if 2.4 gallons of oil is pumped into the cap end port?

3.26 You have a 3.5″ hydraulic cylinder with a 1.5″ diameter piston rod. How far will the rod retract when $\frac{1}{10}$ gallon of oil is pumped into its rod end port?

3.27 You have a 5″ hydraulic cylinder with a 1.85″ diameter piston rod. How far will the rod retract when $\frac{6}{10}$ gallon of oil is pumped into its rod end port?

3.28 You have a 6″ hydraulic cylinder with a 2.35″ diameter piston rod. How far will the rod retract when $\frac{9}{10}$ gallon of oil is pumped into its rod end port?

3.29 You have a 7″ hydraulic cylinder with a 2.35″ diameter piston rod. How far will the rod retract when $\frac{8}{10}$ gallon of oil is pumped into its rod end port?

3.30 At what rate will a piston rod extend when oil is pumped into the cap end port of a 5″ cylinder at 1.75 GPM?

3.31 At what rate will a piston rod extend when oil is pumped into the cap end port of a 2″ cylinder at .25 GPM?

3.32 At what rate will a piston rod extend when oil is pumped into the cap end port of a 3.25″ cylinder at .5 GPM?

3.33 At what rate will a piston rod extend when oil is pumped into the cap end port of a 4.75″ cylinder at 1.2 GPM?

3.34 You have a 3.5″ cylinder with a 1″ piston rod. At what rate will the rod retract if oil is pumped into the rod end port at .85 GPM?

3.35 You have a 5″ cylinder with a 1.5″ piston rod. At what rate will the rod retract if oil is pumped into the rod end port at 1.25 GPM?

3.36 You have a 6.5″ cylinder with a 2″ piston rod. At what rate will the rod retract if oil is pumped into the rod end port at 2.5 GPM?

3.37 You have a 4.35″ cylinder with a 1.2″ piston rod. At what rate will the rod retract if oil is pumped into the rod end port at .75 GPM?

3.38 In how many seconds will the piston rod of a 6″ cylinder extend 4.5″ when oil is pumped into the cap end port at 3.5 GPM?

3.39 In how many seconds will the piston rod of a 5″ cylinder extend 2″ when oil is pumped into the cap end port at 1.5 GPM?

3.40 In how many seconds will the piston rod of a 8″ cylinder extend 1.25″ when oil is pumped into the cap end port at 3.3 GPM?

3.41 In how many seconds will the piston rod of a 2.5″ cylinder extend 3″ when oil is pumped into the cap end port at .75 GPM?

The Basic
Hydraulic System

4

While practical hydraulic systems range from the very simple to the mind-boggling complex, there are certain components which must be included for the apparatus to be practical and functional. In this chapter we will learn what these vital elements are. Later we will see how other devices can be added to improve the system and add to its capabilities. Whatever its purpose, any industrial hydraulic system will have at least these:

1. **A fluid, usually oil**
2. **A tank, or reservoir, to hold a supply of fluid**
3. **Fluid conditioning devices to keep the fluid clean and cool**
4. **A prime mover, usually an electric motor or engine, to drive the pump**
5. **A pump, to cause fluid to flow**
6. **Conductors, usually pipe or tubing, to carry the fluid**
7. **Valving, to control fluid flow, direction, and pressure**
8. **One or more actuators, usually cylinders or hydraulic motors, to do the work**

You should know how each of these elements contributes to the overall operation of a system, and what happens when it fails to do its job. This will be very important when you need to diagnose problems in a circuit. In this chapter we will examine briefly the elements that make up a simple hydraulic circuit, learn how each contributes to its operation, and become familiar with schematic symbols. We will "assemble" the circuit as a series of figures, the first showing only a reservoir and the last showing all the components in place.

HYDRAULIC FLUID

The fluid first used in hydraulic systems was water. Water, however, has several undesirable characteristics which make it unsuitable for use in modern industrial systems. It evaporates, especially when heated, freezes, leaks, rusts parts, and is a poor lubricant. Nearly all modern systems use some form of petroleum oil. This was first used as a hydraulic fluid on the USS *Virginia*, and since that time, many kinds of additives and refinements in processing have greatly improved its effectiveness for this purpose. You need to know a few of the terms used to describe properties of hydraulic fluid.

Viscosity is a measure of oil's **resistance to flow**. It is determined under laboratory conditions by measuring the time required for a specific amount of oil at a specific temperature to flow through a small opening. The higher the viscosity, the longer it takes. We often think of high viscosity oil as being "thick" and low viscosity as being "thin." The **Viscosity Index** is a measure of how much the viscosity **changes** when the oil is heated. This is important only if the system is subjected to high levels of heat. An oil is said to have a high viscosity index if the viscosity changes very little. This is a good quality to have, as the lubricating capability remains constant.

Pour Point is the **lowest temperature** at which a given oil will flow enough to be useful, as determined by set standards. This is important if the system must operate in very cold conditions.

Flash Point is the temperature at which a given oil will give off enough gas to "flash," or "light" briefly if touched with a flame.

Fire Point is the temperature at which a given oil, if touched with a flame, will catch fire and stay lit for at least five seconds. **Auto-ignition Temperature** is the degree of heat at which a given oil will catch fire by itself. This is the principle that drives diesel engines. When the high temperature generated by compression inside the cylinders reaches the auto-ignition temperature of the diesel oil, it ignites and produces power.

The fluid used in any practical hydraulic system must meet two sets of requirements: (1) it must fulfill certain **functions** necessary for the circuit to do the required work; and (2) it must have certain **characteristics** so as to perform these functions effectively without harm to system components.

This hydraulic fluid then performs three important functions:

1. It **transmits force** to the actuators doing the work of the system;
2. It **conducts heat** away from moving parts; and
3. It **lubricates** moving parts to reduce heating, lessen wear, and increase efficiency.

Therefore, certain **characteristics are desired in hydraulic fluid.** It should

1. Be of **appropriate viscosity** for the specific application. If too high ("thick"), it reduces efficiency, increases pressure drop in lines, and generates heat. If too low ("thin"), it tends to leak past seals and will not lubricate;
2. Retain its **lubricating abilities** even when subjected to high heat (viscosity index);
3. Be **free of chemicals** which might damage system components or seals;
4. **Resist "foaming"** when subjected to turbulence or splashing;
5. Be **fire-resistant**, if in an environment where fire is seen as a potential hazard (flash point); and
6. Be **resistant to oxidation**, which converts oil into gum, sludge, and varnish.

Producers of hydraulic fluids use substances called **additives** to achieve these characteristics. Other substances, called **contaminants**, often get into the fluid, cause problems, and must be removed. There are three principal sources of contaminants:

1. Dirt and pieces of material found in components and lines when **first installed.** All incoming parts should be inspected before installation, and a new system should be flushed and the fluid replaced after an initial "run-in" period;
2. Dirt and other foreign particles **entering** the system from the air. A tank should be equipped with an air filter; and
3. Chemicals, gum, sludge, and varnish **generated** by the system in operation, and moisture from **condensation.**

Several kinds of fluids are used in hydraulic systems, and a wide variety of materials are used in seals, valves, conductors, and other components. These are chosen by engineers and designers and tested to ensure that they are **compatible** — that is, they will not damage each other. When servicing hydraulic equipment, take care to use **only** the type of oil and replacement parts specified.

HYDRAULIC RESERVOIRS OR TANKS

Unlike pneumatic systems, which draw air from the atmosphere and vent it back after use, hydraulic systems must have a place to store the fluid used. This place is called a reservoir or tank. It serves functions other than storage, however, and is actually a working part of the circuit. The most common cause of problems occurring in the operation of hydraulic systems is improper maintenance of the tank and the fluid it contains. A hydraulic reservoir or tank serves the following six functions:

1. **Stores** the hydraulic fluid of the system, including some reserve;
2. **Protects** the stored fluid from outside contamination;
3. Provides means by which the **amount** of fluid in the system can be **checked**;
4. Provides means by which fluid can be **added** or **changed** when necessary;
5. **Cools** the fluid as it returns from the actuators; and
6. **Removes contaminants** such as water, dirt, pieces of metal, or chemicals from the fluid.

Tanks vary somewhat in size, construction, and accessories installed depending upon the particular needs of each application. The functions, however, tend to be carried out in much the same way. Figure 4-1 represents a typical hydraulic tank. Most tanks are of welded steel construction, with supports for mounting. This allows for easy access to the **drain plug** in the bottom and also permits cooling air to circulate underneath. The fluid capacity should normally be two to three times the amount of fluid pumped in one minute, to ensure enough reserve, and enough to allow cooling between cycles to keep the temperature below 150° F.

Figure 4-1 Hydraulic reservoir, or tank

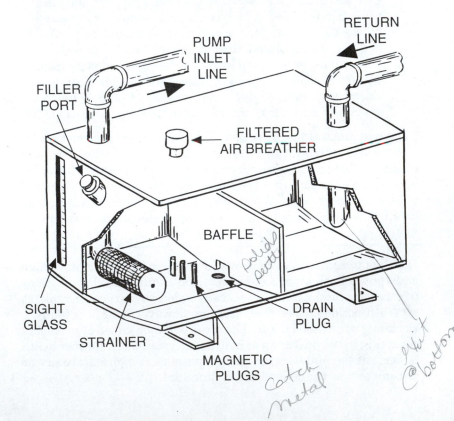

A tank must be **totally enclosed,** and should have a **filtered air breather** to screen out particles from the surrounding atmosphere. When the system is operating, the fluid level is constantly rising and falling, causing air to enter and leave. Tanks located in corrosive or very unclean areas are sometimes pressurized to keep out contaminants. Where acids and other chemicals are present, even more complex means than a simple filter must be used to keep the fluid pure. These two functions performed within the tank, cooling and the elimination of contaminants, address the two major obstacles to efficient, trouble-free system performance.

The simplest and most common device for checking the fluid level is a **sight glass** on the side of the tank. This is checked periodically, and fluid is added as needed. Some systems have automatic sensing devices with a light or horn to alert maintenance personnel of low fluid levels. This provides greater reliability but at higher cost. A **filler port,** with a plug or cover, provides access to the tank for adding fluid. It is constructed and located to keep oil from being contaminated or spilled, resulting in slippery floors or fire hazards. Some tanks are equipped with **magnetic plugs** to attract metal particles. These must be removed and cleaned periodically as part of a regular maintenance program.

A **strainer** blocks the relatively large solid particles from entering the system. This is attached to the **pump inlet line** and may be immersed in the oil near the bottom of the tank. Elsewhere, preferably in the return line, a **filter** is used to remove smaller particles by a process called absorption.

Figure 4-2 Complete hydraulic power unit. *Courtesy of Polypac Systems Center, Continental Hydraulics, Savage, MN*

A **strainer is a fine metal screen which cleans contaminant particles from the system by blocking, or excluding them. Those that stick to the screen are cleaned off later, and the screen is reused.**

A **filter is made of a porous material which allows fluid to flow through, but which traps and holds ("absorbs") very small contaminant particles. When it becomes clogged, it is thrown away and replaced.**

The actual location of both strainer and filter may vary in hydraulic circuits. A strainer located outside the tank, in the pump inlet line, is convenient to inspect, clean, or change, but unless carefully tightened when installed, may allow air into the system. If placed in the tank, below the level of the fluid, it need only be hand-tightened, but is not as convenient to service. A filter located in the pump inlet line between the strainer and pump ensures that small particles of contaminants are kept out of the system, but

Figure 4-3 Contamination by condensation inside the tank

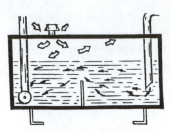

(1) MOIST AIR ENTERS THROUGH AIR BREATHER WHEN SYSTEM IS IN OPERATION.

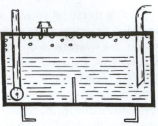

(2) TANK COOLS WHEN SYSTEM IS SHUT DOWN. MOISTURE IN THE AIR CONDENSES INTO WATER DROPLETS ON THE TANK SURFACE.

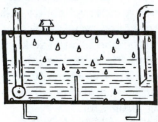

(3) WATER DROPLETS GROW BIG AND HEAVY AND DROP INTO THE HYDRAULIC FLUID.

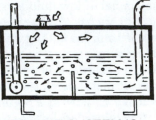

(4) WHEN THE SYSTEM IS AGAIN IN OPERATION, WATER MIXES WITH THE FLUID AND IS CARRIED THROUGH THE SYSTEM.

Figure 4-4 Filters with sintered bronze or stainless steel mesh elements are used for pressures to 5000 psi. *Courtesy of Power Team Div., SPX Corp., Owatonna, MN*

its resistance to flow causes a pressure drop in the fluid being drawn into the pump. Locating the filter in the return line eliminates this problem, but the filtering takes place only after contaminants have already gone through the system. Some may become trapped in the pump, valves, or actuators and never get to the filter. Another source of contamination is **condensation**, which occurs when moist air enters the tank, cools into water droplets, and drops into the fluid (see Figure 4-3).

Contaminant particle size is measured in **microns**. One micron is one-millionth of a meter, or .000039″. The finest-mesh **strainers** are capable of excluding most particles down to about 40 microns, but the most common size removes those between about 200–250 microns. **Filters** are made to absorb particles as small as 1 micron, but 3–5 micron filtration is typical. Figure 4-4 shows one type of filter, and Figure 4-5 gives you an idea of how small these particles are.

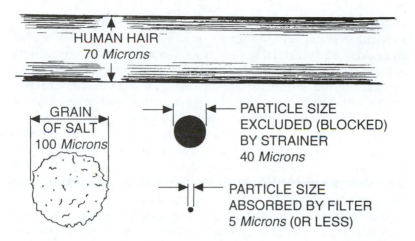

Figure 4-5 Relative size of particles removed by strainers and filters

HUMAN HAIR
70 *Microns*

GRAIN
OF SALT
100 *Microns*

PARTICLE SIZE
EXCLUDED (BLOCKED)
BY STRAINER
40 *Microns*

PARTICLE SIZE
ABSORBED BY FILTER
5 *Microns* (0R LESS)

A wall, called a **baffle**, is used to disrupt the flow between the return line and the pump inlet line. This causes continuous mixing, preventing the development of a narrow stream of oil flowing directly from the return line to the pump inlet. The results are a more **even fluid temperature**, release of any **entrapped air**, and more thorough **settling of contaminants** to the tank bottom. Ideally, the baffle should be no higher than about two-thirds of the fluid depth.

On a schematic diagram of a system, a tank is shown as a "U" shaped symbol with a line representing the pump inlet line going all the way to the bottom. As you can see in Figure 4-6, strainers are usually shown, either in the line or submerged in the tank, but filters are usually omitted. The same symbol is used for either a strainer or filter. Return lines from valves and actuators are usually shown with separate tank symbols to simplify the diagram just as separate "ground" symbols are used on electrical diagrams, but all the symbols represent the same tank.

Figure 4-6 Tank symbols on schematic diagrams

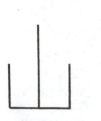

TANK WITH
PUMP INLET LINE.

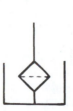

TANK WITH
IMMERSED STRAINER.

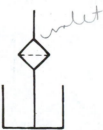

TANK WITH
STRAINER IN
PUMP INLET LINE.

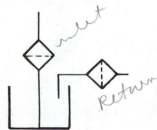

TANK WITH
STRAINER IN
PUMP INLET LINE,
FILTER IN RETURN.

HYDRAULIC PUMPS

A **pump** is a mechanical device which causes **fluid flow** in the system. It is driven by a **motor** or **engine** called a **prime mover**. For nearly all industrial hydraulic systems, the prime mover is an electric motor. This provides a very reliable source of energy at a constant speed. For other applications, such as on construction equipment, the pump is usually driven by a diesel or gasoline engine. The prime mover usually drives the pump through a flexible coupling, so either may be conveniently replaced if necessary.

The most common type of pump is the **external gear pump** shown in Figure 4-7. In its simplest form, this consists of two gears within a housing, one of which is driven by the prime mover and positioned to mesh with the other so that they rotate in opposite directions. The hydraulic fluid in the tank is acted upon by atmospheric pressure (about 14.7 psi at sea level), so when the gears rotate, suction is created at the inlet port of the pump. The fluid is drawn into the pump and carried in the spaces between the gear teeth to the discharge port. As rotation continues, the teeth of the two gears come together leaving no room for the fluid which is then forced out through the discharge port.

It is important to remember that the only pressure on the inlet side of the pump is atmospheric pressure and that the fluid is drawn into the pump by a **reduction** in this pressure, commonly called a **suction** or **vacuum**. If there were no atmospheric pressure, as in outer space, the pump would not work. Rotation of the gears causes **fluid flow**; no pressure exists until there is resistance to

Figure 4-7 Operation of the external gear pump

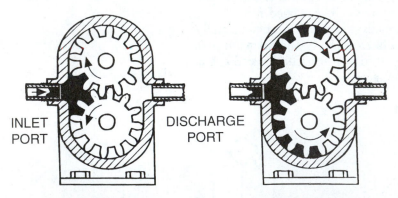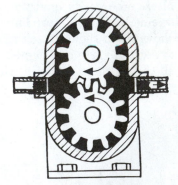

INLET PORT   DISCHARGE PORT

ATMOSPHERIC PRESSURE IN THE TANK FORCES FLUID INTO THE INLET PORT.

FLUID IS CARRIED THROUGH THE PUMP IN THE SPACES BETWEEN GEAR TEETH.

AS TEETH MESH TOGETHER ON THE DISCHARGE SIDE, FLUID IS FORCED OUT THROUGH THE DISCHARGE PORT.

this flow. However, the gear teeth **causing** the flow are capable of withstanding high pressure. When that flow is blocked or resisted, system pressure builds.

Most pumps used in industrial hydraulic systems, including the gear pump just described, are **positive-displacement** pumps. This means that when the pump is running, the fluid **must be displaced** or **moved,** or something will break or the prime mover will stall. A **non-positive displacement pump** just moves fluid but cannot build up much pressure. Fans, blowers, and propellers are capable of non-positive displacement because if the flow is blocked, they can continue to run and generate turbulence without moving fluid.

Figure 4-8 shows the standard symbols for a hydraulic pump and its prime mover (motor) as they appear on a schematic diagram. The black triangle on the pump symbol identifies the discharge port. Other types of positive displacement pumps will be covered in Chapter 6.

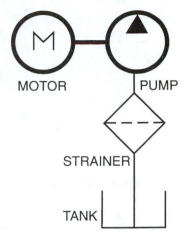

Figure 4-8 System schematic showing location of the pump and motor

PRESSURE RELIEF VALVES

Since fluid from a positive-displacement pump must flow continuously whenever the pump is running, it must have somewhere to go when not being used by the actuators. Normally it is directed back to the tank through a pressure relief valve.

A pressure relief valve has two functions:

1. **Set a maximum operating pressure level for the system; and**
2. **Protect system components such as the pump, valves, and lines from overpressure.**

The pressure relief valve acts much like the main circuit breaker in your home electrical system. Should a short circuit develop in your house wiring, power is interrupted before reaching any internal circuits. In the hydraulic system, the pressure relief valve is the first system component following the pump. If high pressure develops in the system, the fluid drains back to the tank, dropping the pressure to the valve setting.

Figure 4-9 illustrates one type of pressure relief valve. It is located in the line on the **discharge** side of the pump, with a drain line to carry fluid back to the tank. The pressure level is set by means of an adjustment screw and compression spring. The ball remains seated in normal operation, as system pressure is not strong enough to compress the spring. If system pressure **exceeds** the set level, the spring is compressed and the **ball** is **lifted** from its seat, allowing fluid to return to the tank through the drain line. Pressure settings on the pressure relief valve should be made only

Figure 4-9 Operation of a pressure relief valve

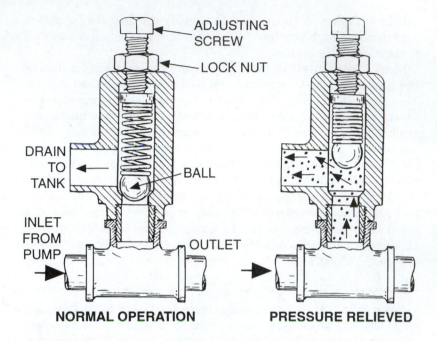

ADJUSTING SCREW

LOCK NUT

DRAIN TO TANK

BALL

INLET FROM PUMP

OUTLET

NORMAL OPERATION   **PRESSURE RELIEVED**

by system installers, technicians, or maintenance mechanics—**not** by system operators. When possible, this valve should be located or constructed in such a way that it is not likely to be tampered with. The **security** of the entire system and the **safety** of workers depend on the protection it provides.

Figure 4-10 shows the standard symbol for a pressure relief valve as it would appear on a schematic diagram. Hydraulic symbols are designed to help you visualize their functions. Here, you

Figure 4-10 System schematic showing location of the pressure relief valve

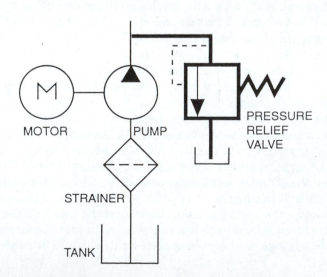

MOTOR   PUMP

PRESSURE RELIEF VALVE

STRAINER

TANK

imagine pressure building in the dashed line (called a "pilot") and moving the arrow to the right. As it centers, visualize fluid being able to flow through it to the tank. Remember, there is only one tank and a separate tank symbol is shown here only to simplify the drawing.

A single pump may provide fluid for the operation of several actuators, which may require **different** hydraulic pressures for their operation. Therefore, while the pressure relief valve sets a maximum pressure for the system **as a whole,** there may exist a need for individual **regulated** or **controlled** pressures in **individual** lines. A **pressure reducing valve** serves this purpose.

Figure 4-11 illustrates one type of pressure reducing valve. It is placed in a **branch** line serving one or more actuators that require the **same** level of hydraulic **pressure.** The pressure level is maintained by the adjustment screw and compression spring. Notice that it differs from the pressure relief valve in that a **spool** is used in place of a **ball.** In **normal** operation, fluid passes freely **through**

PRESSURE REDUCING VALVES

Figure 4-11 Operation of a pressure reducing valve

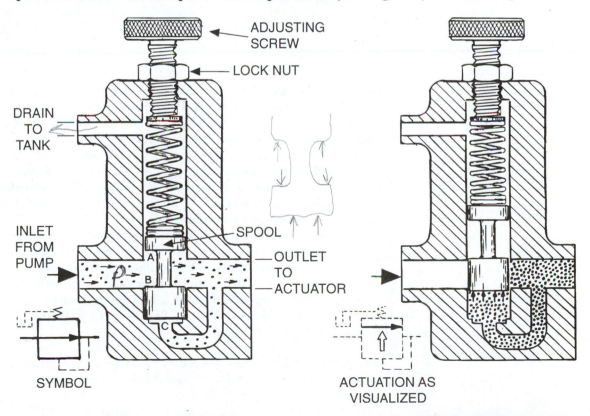

ADJUSTING SCREW

LOCK NUT

DRAIN TO TANK

INLET FROM PUMP

SPOOL

OUTLET TO ACTUATOR

SYMBOL

ACTUATION AS VISUALIZED

NORMAL OPERATION

PRESSURE REGULATED

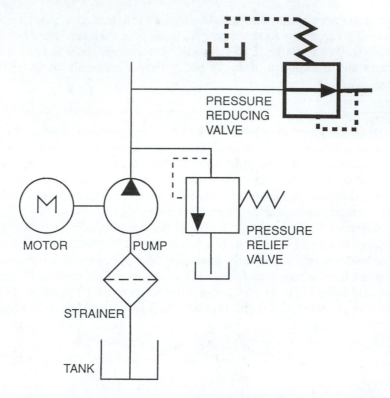

Figure 4-12 System schematic showing location of the pressure reducing valve

the valve to the **actuator;** and within the normal operating range, pressure is exerted on surfaces **A, B,** and **C** of the spool. Surfaces **A** and **B** have equal areas, so the **forces** exerted on them by the pressure of the fluid are **equal** and **opposite.** Thus, they cancel each other. Note the line leading back from the outlet to surface **C.** This is called a **pilot** line because its function is to carry fluid to **control the valve** rather than to power an actuator. The pressure of fluid at **C** exerts an **upward** force on the spool which is resisted by the spring, holding the spool seated.

Should line pressure rise to a level above the setting, the force exerted at C lifts the spool and **blocks** further flow to the actuator. Note here the difference between this valve and the pressure relief valve: Fluid is **not directed back to the tank,** but is **kept in the lines** so it may be used for operating other actuators in other branches. A drain line is provided so that any fluid leaking past the spool will be taken back to the tank.

The purpose of the pressure reducing valve is to regulate, or control the pressure for an individual circuit branch or actuator.

Figure 4-12 shows the standard symbol for a pressure reducing valve as it would appear on a schematic diagram. Notice that it is located in a **branch** line, and that the **main** hydraulic line extends upward to **other branches** and actuators.

For a hydraulic system to do the work for which it was assembled, fluid must be delivered to the actuators, at the proper ports, at the appropriate time. This is the function of a **directional control valve.** The simplest type is a **check valve,** which consists of a single passageway and allows flow in one direction only. No check valve is used in the circuit being examined here, but various types will be shown and explained in Chapter 7. A directional control valve may have several passageways. The most common types have a moving part called a **spool,** whose position determines where fluid will flow.

A way is a route which fluid may take through a valve. A check valve has only one way because flow can be routed in only one direction.

A port is an opening through which ways are connected with lines leaving the valve. A way may connect with more than one port.

A position is a specific setting, or location, of a spool which determines through which ways fluid will flow, if any.

Directional control valves are described by specifying (1) the number of **spool positions,** (2) the number of **ways,** (3) the means by which the valve is **operated,** (4) the means by which the spool is returned to **normal,** and (5) for three-position valves, whether **open** or **closed center.** Three-position directional control valves will be discussed in Chapter 7. Figure 4-13 shows typical construction of a **two-position** directional control valve.

Until the pushbutton is pressed, the spool is positioned to the right, in the **normal,** or **nonactuated** position, by the force of the compression spring. Fluid **from** the cylinder is free to flow **into** the CYL port, through the valve, out **PORT T,** and back to the tank. When the pushbutton is **pressed,** the valve shifts to the **actuated** position. Now, fluid from the **pump** flows into **PORT P,** through the valve, and out the CYL port to the cylinder. In this illustration, the piston rod would be retracted by either a load or a retraction spring. In others, a directional control valve may be used to channel fluid flow to both extend and retract the piston rod.

Figure 4-14 shows the relationship between the schematic symbol for this valve and its actual operation. Imagine the symbol and the valve as a two-part box. On the left side of each part are two ports, and on the right, only one. The **upper** part of the box is

Figure 4-13 Two-position, three-way, pushbutton-operated, spring-return directional control valve

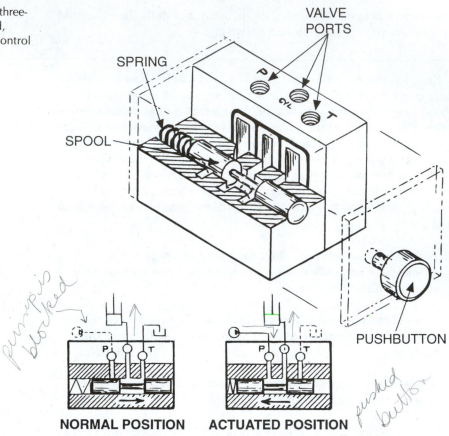

NORMAL POSITION   **ACTUATED POSITION**

Figure 4-14 Interpretation of schematic valve symbol

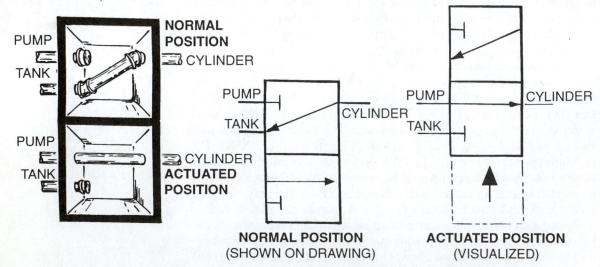

NORMAL POSITION
(SHOWN ON DRAWING)

ACTUATED POSITION
(VISUALIZED)

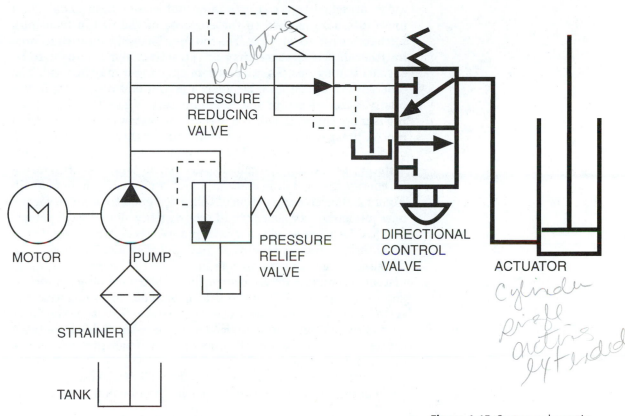

Regulating

Cylinder single acting extended

Figure 4-15 System schematic showing location of directional control valve and actuator

piped for fluid flow in the **normal** position. The input line from the **pump** is blocked, while fluid can flow from the **actuator** directly to the **tank**. The **lower** part of the box is piped for fluid flow in the **actuated** position. Here the line to the **tank** is blocked, while fluid can flow from the **pump** to the **actuator**. On a schematic, the symbol is usually shown with the valve in the **normal**, or **nonactuated** position. To interpret fluid flow, you need to **imagine** the symbol on the drawing as being able to **shift** to its actuated position. Figure 4-15 shows the placement of a directional control valve and its connections to an actuator in a circuit.

A hydraulic actuator is a device which converts fluid energy into mechanical force and motion.

ACTUATORS

Two **basic** types of actuators are used in industrial hydraulic systems, with each type having several **variations**. **Cylinders** are used to produce motion in a **straight line**, which we call **linear** motion. A movable element called a **piston** is fitted inside the cylinder, attached to a rod extending out **one** or **both** ends. Fluid under pressure applied to the surfaces of the piston transmits force to the rod to do work. The rod may be used to transmit force in either direction—**extending** or **retracting**. When force is to be exerted in only **one** direction, a **ram** or **plunger** may be used. The difference in appearance between a **piston** and a **ram** lies in the relative sizes of **piston** and **rod** diameters. When the **area** of the **rod** is more than **one-half** the area of the **piston face**, it is called a **ram**. A **plunger** is essentially a cylindrically shaped movable piece used in a cylinder or bored hole to convert pressure into force.

Rotary actuators, also called **hydraulic motors,** convert hydraulic energy into **rotating** motion about an axis. As is evident in Figure 4-16, the construction of those designed to produce **continuous** motion is very similar to the construction of hydraulic **pumps.** Just as there are several types of pumps, there are several types of **motors,** including **gear, vane,** and **piston** models. In each case, the **shaft,** which in a pump was the **driver** of the gears, vanes, or pistons, becomes a **driven** element in a **rotary actuator,** or **motor,** shown in Figure 4-17. In the gear type, fluid forced against the **teeth** of the two gears causes them to **rotate.** Note that the lower gear is **more** than just an **idler**—the force of the fluid on its teeth causes it to rotate against the upper gear and **help** generate **torque** on the output shaft.

If less than 360° of rotation is needed, a **limited rotation** actuator is used. This is much simpler in **construction** and therefore less

Figure 4-16 Hydraulic motor

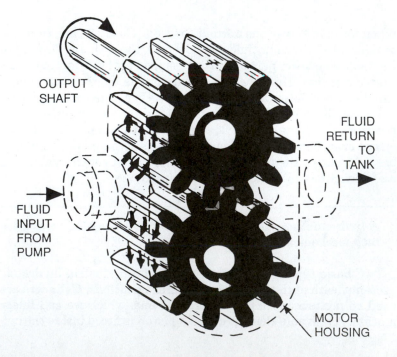

OUTPUT
SHAFT

FLUID
RETURN
TO
TANK

FLUID
INPUT
FROM
PUMP

MOTOR
HOUSING

Figure 4-17 Hydraulic rotary actuator. *Courtesy of Parker-Hannifin Corp., Fluidpower Group*

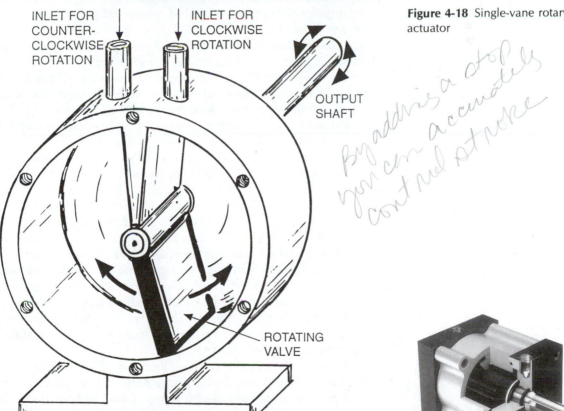

INLET FOR COUNTER-CLOCKWISE ROTATION

INLET FOR CLOCKWISE ROTATION

OUTPUT SHAFT

ROTATING VALVE

Figure 4-18 Single-vane rotary actuator

By adding a stop you can accurately control stroke

Figure 4-19 Typical vane type rotary actuator construction. *Courtesy of Parker-Hannifin Corp., Fluidpower Group*

expensive, is readily **reversible,** and is capable of starting, stopping, and reversing motion much **faster** than either hydraulic or electric motors. There are several types of rotary actuators, with single and multiple vanes providing variations in torque and degree of rotation. Figures 4-18 and 4-19 illustrate two types of rotary actuators.

CHAPTER SUMMARY

While there is considerable variation in the size and complexity of industrial hydraulic systems, any practical circuit will have at least these eight components: (1) **fluid**, (2) **a tank**, (3) **fluid conditioning devices**, (4) **a prime mover**, (5) **a pump**, (6) **conductors**, (7) **valving**, and (8) **actuators**. Water is unsuitable as a hydraulic fluid because **it evaporates, freezes, leaks, rusts parts**, and is a **poor lubricant**.

Viscosity is a measure of oil's **resistance to flow**. The **viscosity index** is a measure of the **viscosity change** when a given oil is **heated**. **Pour Point** is the **lowest temperature** at which a given oil will flow enough to be useful. **Flash Point** is the temperature at which oil will **flash**, or light briefly if touched with a flame. **Fire Point** is the temperature at which a given oil, if touched with a flame, will catch fire and stay lit for at least five seconds. The **auto-ignition temperature** is the level of heat at which a given oil will catch fire by itself.

A hydraulic fluid must fulfill **three functions**: (1) **transmit force**, (2) **conduct heat**, and (3) **lubricate**. A hydraulic fluid should have these **six characteristics**: (1) **appropriate viscosity**, (2) **lubricating ability**, (3) **freedom from damaging chemicals**, (4) **resistance to foaming**, (5) **fire-resistance**, and (6) **oxidation-resistance**.

Additives are used to achieve desired characteristics in hydraulic fluid. **Contaminants** in hydraulic fluid are a major source of problems in system operation. The three principal **sources** of contaminants are (1) **newly installed components**, (2) **air entering the tank**, and (3) **system operation**, generating chemicals, gum, sludge, varnish, and moisture. The **functions** of a **tank** are: (1) **store fluid**, (2) **protect fluid**, (3) provide means for **checking fluid level**, (4) provide means for **adding fluid**, (5) **cooling**, and (6) **remove contaminants**. **Large** contaminant particles are **blocked** from entering the system by a **strainer**. **Small** particles are removed by being **absorbed** in the material of a **filter**. Contaminant **particle size** is measured in **microns**. A wall in the tank, called a **baffle**, causes mixing in the flow between the return line and the pump inlet line. This results in a more **even fluid temperature** and more thorough **settling of contaminants** to the tank bottom.

The purpose of a **pump** is to cause **fluid flow**. The motor or engine that drives a pump is called a **prime mover**. The most common type of pump is the **external gear pump**. Two gears within a housing carry fluid between the gear teeth from the **inlet port** to the **discharge port**. Fluid is forced into the **inlet port** of a pump by **atmospheric pressure** when suction is created inside the pump. Most pumps used in industrial hydraulic systems are classified as **positive-displacement** pumps. This means that fluid **must be moved** when the pump is running.

The two functions of a **pressure relief valve** are to (1) **set a maximum operating pressure** and (2) **protect components from over-pressure**. The purpose of a **pressure reducing valve** is to **regulate** or **limit** pressure for an individual circuit branch or actuator. The purpose of a **directional control valve** is to **deliver** fluid to an actuator at the **proper port** at the **appropriate time**. A **way** is a flow path through a valve. A **port** is an opening in the valve body. A **position** is a specific setting of a spool which determines flow paths. **Directional control valves** are described by specifying the number of **ways**, the number of **positions**, the means by which the valve is **operated**, and, for three-way valves, whether **open** or **closed center**. Finally, an **actuator** is a device which converts **fluid energy** into **mechanical force** and **motion**.

PROBLEMS

4.1 List the eight components, parts, or devices that are found in any industrial hydraulic system. pg 61

4.2 List five reasons why water is not suitable for use as a hydraulic fluid. pg 63

4.3 Explain, in your own words, the term "viscosity" and how, in a laboratory, this is determined for a given sample of fluid. pg 62

4.4 Explain, in your own words, the term "viscosity index" and why this may be an important factor in the selection of a hydraulic fluid for a particular application. pg 62

4.5 Explain, in your own words, the term "pour point" and why this may be an important factor in the selection of a hydraulic fluid for a particular application. pg 62

4.6 Explain, in your own words, the difference between "flash point," "fire point," and "auto-ignition temperature" with regard to oil. pg 62

4.7 List three functions performed by the fluid itself in a hydraulic system. pg 62

4.8 List four characteristics which are desirable to have in a hydraulic fluid. pg 63

4.9 Explain briefly three ways in which hydraulic fluid may become contaminated. pg 63

4.10 What do we mean when we say that a hydraulic fluid and the materials used in seals, valves, and other components must be "compatible?" pg 63

4.11 List four functions served by the tank, in addition to storing fluid while it is not being used. pg 64

4.12 Explain briefly the process by which hydraulic fluid becomes contaminated from condensation. pg 66

4.13 Explain the difference between a strainer and a filter, and the way in which each removes particle contaminants from hydraulic fluid. pg 65

4.14 Describe a "baffle" and explain briefly its purpose in a tank. pg 67

4.15 Draw the symbol for each of the following as it would appear on a schematic diagram:
   (a) Tank 67
   (b) Strainer 67
   (c) Motor 69
   (d) Pump 69
   (e) Pressure Relief Valve 70
   (f) Pressure Reducing Valve 72
   (g) Cylinder (Actuator) 75

4.16 What is the purpose of the pump in a hydraulic system? cause fluid flow

4.17 What causes fluid to move from the tank to the inlet port of the pump in a hydraulic system? at mospheric pressure

4.18 What causes pressure to develop in a hydraulic system when it is in operation? a reduction in atmosph

4.19 Explain briefly the function of a pressure relief valve. pg 69

4.20 Explain briefly the function of a pressure reducing valve. pg 71

4.21 Explain briefly the function of a directional control valve. 73

4.22 What are the two major obstacles to good system performance? b
   (a) improper oil viscosity and low system pressure
   (b) fluid contamination and heat
   (c) metal fatigue and improper valve operation
   (d) incorrect pressure relief valve settings and clogged filters

4.23 Hydraulic pressure in the system is caused by which one of the following? d
   (a) the pump
   (b) a pressure relief valve
   (c) a force acting upon an area
   (d) resistance to fluid flow

4.24 The purpose of a pressure relief valve is to do which one of the following?
   (a) limit fluid flow to the actuators
   (b) maintain a specific pressure in the system   d
   (c) relieve fluid pressure in a cylinder
   (d) limit maximum pressure in the system

4.25 The purpose of a pressure reducing valve is to do which one of the following?
   (a) limit pressure in a branch line
   (b) limit pressure at the pump discharge port   a
   (c) protect components throughout the system
   (d) keep pressure from dropping to zero

4.26 The purpose of a directional control valve is to do which one of the following?
   (a) prevent fluid from flowing backward in the lines
   (b) direct fluid from the tank to the pump   c
   (c) direct flow to the actuators at proper ports, at appropriate times
   (d) direct fluid from the pump to the tank when pressure becomes too high for the system

4.27 What is the difference between a hydraulic cylinder and a hydraulic ram?

(a) A ram is meant to be used for higher forces

(b) A cylinder is longer

(c) The rod of a cylinder usually will retract at a faster rate

(d) The rod of a ram has a cross-sectional area greater than half that of the piston face.

System Calculations

The operation of any hydraulic system requires energy. While scientists tell us we can neither **create** nor **destroy** energy, the operation of any working system nearly always requires that energy be **released** or **converted** into a specific **form**. Energy stored in coal, oil, or wood is released in the form of **heat**, by burning. Energy stored in a battery is released in the form of **electricity** when made part of a circuit. Energy stored in a compressed spring, when released, is **mechanical** energy capable of exerting force and moving objects.

Electrical energy delivered to homes and factories is converted into **mechanical** energy by electric motors, **heat** by stoves and burners, and **light** by lamps.

Energy is defined as the ability to do work.

While energy exists in several forms, we are concerned here with only three—**mechanical, electrical,** and **heat.** Hydraulic systems are **mechanical** because they exert force and cause movement. However, the energy **source** for hydraulic systems is most often an **electric** motor. Energy **lost** or wasted in a hydraulic system is usually converted into **heat.**

Energy is measured by the **amount** of work it can accomplish. **Mechanical** energy is measured in **horsepower. Electrical** energy is measured in **watts. Thermal,** or **heat** energy is expressed in **British Thermal Units (BTUs) per hour.**

MECHANICAL ENERGY

One horsepower is defined as the amount of energy required to lift 33,000 lbs. one foot in one minute.

Most problems in hydraulics use time intervals measured in seconds rather than minutes. Dividing by 60 we find that 33,000 lbs. in one minute (60 seconds) is equal to 550 lbs. in one second. For convenience then, we can use as the definition of horsepower:

One horsepower is the amount of energy required to lift 550 lbs. one foot in one second (see Figure 5-1).

Given the force needed, the distance to be moved, and the time limit, we find the horsepower required for a specific application using the formula

$$HP = \frac{(F)(d)}{550(t)}$$

where

$$HP = \text{Horsepower}$$
$$F = \text{Force in lbs.}$$
$$d = \text{distance in ft.}$$
$$t = \text{time in seconds.}$$

Example: What horsepower is needed to lift 275 lbs. 8 ft. in 2 seconds?

$$HP = \frac{(F)(d)}{550(t)}$$

Figure 5-1 One horsepower

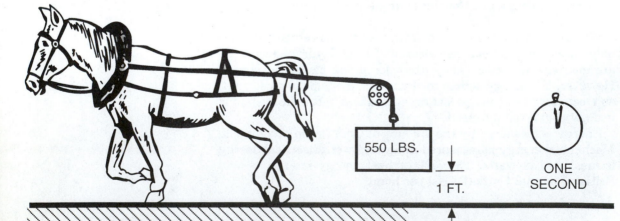

550 LBS.

1 FT.

ONE SECOND

$$HP = \frac{(275)(8)}{550(2)}$$

$$HP = \frac{2200}{1100}$$

$$HP = 2 \text{ horsepower (Ans.).}$$

If the time is given in **minutes,** multiply the minutes figure by **60** to convert to **seconds** as required in this formula.

Example: 3 minutes times 60 equals 180 seconds. Use 180 for t in the horsepower formula.

We need a basic understanding of electricity because electric **motors** are often used to drive hydraulic **pumps.** Electrical energy is produced when electrons are made to move through a material called a **conductor,** usually a wire, by an **uneven distribution** of these electrons. This unbalance is called **voltage.** We can think of this movement of electrons as being like fluid flow in a hydraulic system. The voltage source, then, corresponds to a hydraulic pump. A **pump** moves **fluid. Voltage** moves **amperes,** or "amps," of **current.** Figure 5-2 shows how the parts of an electrical circuit compare with those of a hydraulic system.

ELECTRICAL ENERGY

The unit of electrical energy is the **watt.**

One watt is the amount of energy released when 1 volt is applied to a resistance of 1 ohm, producing a current of 1 ampere ("amp").

The amount of current flow into an electrical device, at a given voltage, depends upon the amount of work the device is required to do. The device is said to "draw" current. The formula for finding the energy input to a motor is

$$W = VA$$

where

W = Watts input
V = Voltage applied
A = Amps of current.

Example: How much electrical energy is used in a 220 volt motor drawing 3 amps of current?

ENERGY SOURCE
PUMP

CONTROL
VALVE

ENERGY USER
ACTUATOR

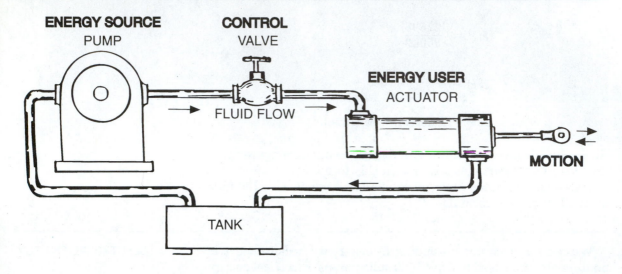

FLUID FLOW

MOTION

TANK

ENERGY SOURCE
VOLTAGE
GENERATOR

CONTROL
SWITCH

CURRENT FLOW

ENERGY USER
ELECTRIC
MOTOR

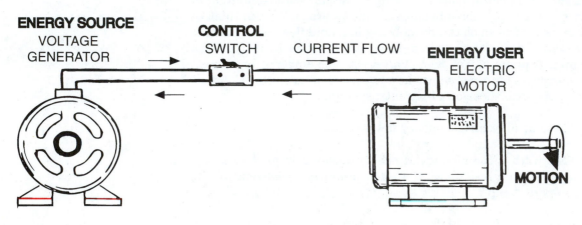

MOTION

Figure 5-2 Comparison between electrical and hydraulic circuits

$$W = VA$$
$$W = (220)(3)$$
$$W = 660 \text{ watts (Ans.)}.$$

If electrical energy were converted into mechanical energy, with no loss, it would take **746 watts** to lift 550 lbs. one foot in one second. Figure 5-3 illustrates how this conversion takes place in an electric motor. We therefore have a way of converting electrical energy into horsepower:

746 Watts equals one horsepower.

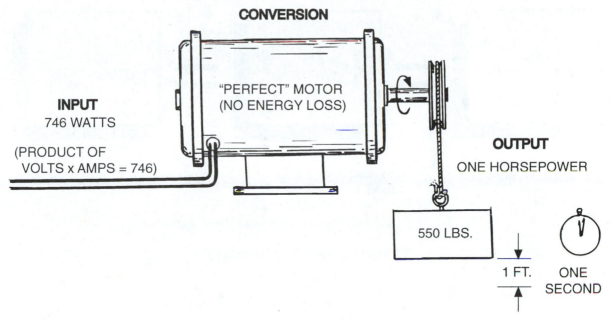

CONVERSION

INPUT
746 WATTS

(PRODUCT OF
VOLTS x AMPS = 746)

"PERFECT" MOTOR
(NO ENERGY LOSS)

OUTPUT
ONE HORSEPOWER

550 LBS.

1 FT.

ONE
SECOND

Figure 5-3 Electrical energy
converted to mechanical energy
by a motor

This is expressed in a formula as

$$HP = \frac{W}{746}$$

where

$$HP = \text{Horsepower}$$
$$W = \text{Watts.}$$

Heat is a form of energy. It can do work, as we see in steam engines. Cold is a term indicating the relative **absence** of this form of energy.

One British Thermal Unit (BTU) is the amount of heat required to increase the temperature of one pound of water by one degree Fahrenheit.

This is actually a small amount of energy. It would take about 74 BTUs to heat one cup of water from room temperature (70°F) to boiling (212°F), if it were possible to do so with no loss of energy.
To convert heat energy to horsepower, we must include a **time**

**HEAT, OR
THERMAL ENERGY**

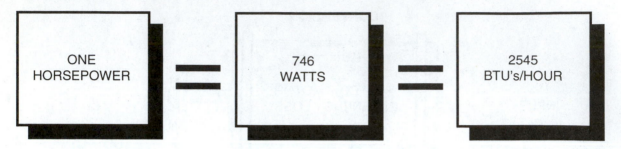

Figure 5-4 Equal values of mechanical, electrical, and thermal energy

limit, and since a BTU is such a small amount of energy, an **hour** is used.

> **2545 BTUs per hour equals one horsepower.**

This is expressed in a formula as

$$HP = \frac{E}{2545}$$

or

$$HP = .00039\ E$$

where

$$HP = \text{Horsepower}$$
$$E = \text{Heat Energy, BTU/Hr.}$$

As Figure 5-4 shows, we have now established equal values for energy in each of the three forms with which we are concerned.

EFFICIENCY

When energy is released or converted in an actual system, part of it is always changed into an unwanted form. Some of the **electrical** energy in a motor or light bulb is changed into unwanted **heat**. In an engine, some of the **chemical** energy in the fuel is also changed into unwanted heat. When fluid flowing through pipes causes them to vibrate, unwanted **mechanical** energy is generated. Remember: No energy is actually **destroyed**. Energy said to be "lost" is only changed into an **unwanted form**.

The relative amount of energy converted which is **useful** is a measure of the **efficiency** of the system.

> **Efficiency is a measure of the amount of system energy which is useful, as compared with the total energy input, and is expressed as a percentage.**

We calculate the efficiency of a device or system using the formula

$$\text{Efficiency} = \frac{\text{Output}}{\text{Input}} \times 100.$$

Multiplying by 100 simply converts the decimal fraction into a percentage. Move the decimal point two places to the right and add the % sign.

Example: What is the efficiency of a machine whose energy input is 8 horsepower and whose output is 6 horsepower?

$$\text{Efficiency} = \frac{\text{Output}}{\text{Input}} \times 100$$

$$\text{Efficiency} = \frac{6}{8} \times 100$$

$$\text{Efficiency} = .75 \times 100$$

$$\text{Efficiency} = 75\% \text{ (Ans.)}$$

In this book, we will calculate answers to the nearest whole percentage. When the calculation yields an answer to more than two decimal places, round off the figure to the nearest whole percentage. If the third figure is a "5," and none follow, round off to the **even** figure. For example:

.824 becomes 82%
.826 becomes 83%
.825 becomes 82%
.835 becomes 84%

When calculating the efficiency of a device or system, the energy values must be in the **same units**. Since the useful output of hydraulic devices is mechanical, we will convert watts (electrical energy) or BTUs/hr. (heat energy) into **horsepower** as the first step.

Example: What is the efficiency of an electric motor that uses 900 watts of energy to produce 1.0 horsepower?

$$HP = \frac{W}{746}$$

$$HP = \frac{900}{746}$$

$$HP = 1.21 \; HP \text{ input}$$

$$\text{Efficiency} = \frac{\text{Output}}{\text{Input}} \times 100$$

$$\text{Efficiency} = \frac{1.0 \, HP}{1.21 \, HP} \times 100.$$

$$\text{Efficiency} = .826 \times 100$$

$$\text{Efficiency} = 83\% \, (\text{Ans.}).$$

If given the voltage input to a motor and the amps it draws, first solve for watts using

$$W = VA,$$

then convert the watts into horsepower. Divide the **output** horsepower by the **input** horsepower to find the motor's efficiency.

Example: What is the efficiency of a 220 volt electric motor that draws 4.8 amps and produces 1.2 horsepower?

$$W = VA \quad \text{Volts} \times \text{current}$$

$$W = (220)(4.8)$$

$$W = 1056 \text{ watts input}$$

$$HP = \frac{W}{746}$$

$$HP = \frac{1056}{746}$$

$$HP = 1.42 \, HP \text{ input}$$

$$\text{Efficiency} = \frac{\text{Output}}{\text{Input}} \times 100$$

$$\text{Efficiency} = \frac{1.2 \, HP}{1.42 \, HP} \times 100$$

$$\text{Efficiency} = .845 \times 100$$

$$\text{Efficiency} = 84\% \, (\text{Ans.}).$$

Two other forms of the efficiency formula may be used in solving problems, depending upon the information given:

$$\text{Output} = (\text{Efficiency})(\text{Input})$$

$$\text{Input} = \frac{\text{Output}}{\text{Efficiency}}.$$

Example: You have a 220 volt electric motor which draws 8.0 amps and operates at 95% efficiency. What is the horsepower output of this motor?

$$W = VA$$

$$W = (220)(8.0)$$

$$W = 1760 \text{ watts input}$$

$$HP = \frac{W}{746} \quad 1570$$

$$HP = \frac{1760}{746}$$

$$HP = 2.36 \; HP \text{ input}$$

$$\text{Output} = (\text{Efficiency})(\text{Input})$$

$$\text{Output} = (.95)(2.36)$$

$$\text{Output} = 2.24 \; HP \text{ (Ans.)}.$$

Example: You are assembling a system whose actuators will require 2.7 horsepower to operate. If the efficiency of the pump is 85%, what must be its input horsepower?

$$\text{Input} = \frac{\text{Output}}{\text{Efficiency}}$$

$$\text{Input} = \frac{2.7 \; HP}{.85}$$

$$\text{Input} = 3.18 \; HP \text{ (Ans.)}.$$

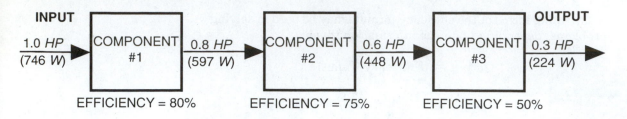

INPUT 1.0 HP (746 W) → COMPONENT #1 0.8 HP (597 W) → COMPONENT #2 0.6 HP (448 W) → COMPONENT #3 0.3 HP (224 W) → **OUTPUT**

EFFICIENCY = 80%   EFFICIENCY = 75%   EFFICIENCY = 50%

$$(.80)\,(.75)\,(.50) = .30 \quad 30\% \text{ of } 1.0 \text{ HP} = 0.3 \text{ HP}$$

Figure 5-5 Efficiency of components in combination

When two or more devices or system components such as an electric motor and a hydraulic pump operate in a **series**, the overall efficiency of the combination is found by **multiplying** the efficiencies of all the components. Figure 5-5 shows how the energy output of each device is further decreased by the one that follows.

Example: You have a hydraulic system in which an electric motor having an efficiency of 90% drives a pump which has an efficiency of 80%. What is the efficiency of the combination?

$$\text{Let } E_m = \text{Efficiency of the motor}$$
$$\text{Let } E_p = \text{Efficiency of the pump}$$
$$\text{Let } E_c = \text{Efficiency of the combination}$$
$$E_c = (E_m)(E_p)$$
$$E_c = (.90)(.80)$$
$$E_c = .72, \text{ or } 72\% \text{ (Ans.)}.$$

CYLINDER HORSEPOWER

The amount of energy needed for a cylinder to function as intended is called cylinder horsepower.

We have already learned the basic formula for **mechanical** horsepower. Now we will develop a modification of this formula for finding **cylinder horsepower**. We begin with the **mechanical** formula

$$HP = \frac{(F)(d)}{550t}$$

where

$$F = \text{Force in lbs.}$$
$$d = \text{distance in ft.}$$
$$t = \text{time in seconds.}$$

The length of piston rod travel, called the **stroke**, is normally given in inches. In the formula, d is in feet and the figure is $\frac{1}{12}$ of what it would be if given in inches. Using S to represent rod travel in **inches**, then:

$$S/12 = d.$$

Example:

60 in. stroke/12 = 5 ft. stroke

Substituting $S/12$ for d in the formula, you get

$$HP = \frac{(F)(S/12)}{550t}.$$

Now multiply both the numerator (upper figure) and the denominator (lower figure) of the fraction by **12** to simplify the compound fraction:

$$HP = \frac{(12)(F)(S/12)}{(12)(550)(t)}.$$

This results in a formula for finding **cylinder horsepower:**

$$HP_{cyl} = \frac{(F)(S)}{6600t}$$

where

$$HP_{cyl} = \text{cylinder horsepower}$$
$$F = \text{Force required}$$
$$S = \text{Stroke length, in inches}$$
$$t = \text{time, in seconds.}$$

Example: What horsepower is required by a hydraulic cylinder to move a load 10 inches in 2 seconds with a force of 3000 lbs.?

$$HP_{cyl} = \frac{(3000)(10)}{6600(2)}$$

$$HP_{cyl} = 2.27 \ HP \ \text{(Ans.)}.$$

This formula assumes an "ideal" system in which there are no energy losses due to friction. Experience has shown that **increasing** the calculated horsepower by **25%** provides an **adequate allowance** for most applications. To do this, multiply the calculated figure by 125%, or 1.25:

$$(1.25)(2.27) = 2.84 \; HP.$$

A 3.0 horsepower motor would be an appropriate choice.

FLUID HORSEPOWER

The cylinder horsepower formula is derived directly from the basic formula for mechanical horsepower, and is useful in determining power requirements for **simple** systems. Another approach is more convenient when we need to analyze the needs of a more complex installation involving several actuators, perhaps of different sizes, or operating at varying times. Whereas the preceding formula focused upon the **output** requirement (force, stroke length, and time), this alternate method is based upon the actual maximum **fluid flow** and **pressure** experienced by a system in operation.

When a piston rod extends, the force exerted depends upon the fluid **pressure** and the piston **area**.

$$F = PA$$

We can therefore substitute in the cylinder horsepower formula:

$$HP_{cyl} = \frac{(F)(S)}{6600t}$$

$$HP_{cyl} = \frac{(P)(A)(S)}{6600t}.$$

When a piston moves within a cylinder, the **volume** of oil moved depends upon the **area** of the piston face and the length of the **stroke.**

$$V = (A)(S)$$

We can therefore substitute V for AS in the formula

$$HP_{cyl} = \frac{(P)(A)(S)}{6600t}$$

$$HP_{cyl} = \frac{(P)(V)}{6600t}.$$

This can also be expressed as:

$$HP_{cyl} = \frac{(P)}{6600} \frac{(V)}{(t)}.$$

Since V is in cubic inches and t is in seconds, V/t represents a value in **cubic inches per second**. Pump outputs are given in **gallons** per **minute**. To convert, multiply by **60** to get cubic inches per **minute**, then divide by **231** to convert cubic inches to **gallons**.

$$\frac{V}{t} = \text{cubic inches per second}$$

$$\frac{60}{231} \times \frac{V}{t} = \text{gallons per minute}$$

Using Q to represent **gallons per minute**:

$$Q = \frac{60}{231} \frac{V}{t}$$

$$Q = .2597 \frac{V}{t}$$

$$\frac{Q}{.2597} = \frac{V}{t}.$$

Going back to the formula and substituting,

$$HP = \frac{P}{6600} \frac{V}{t},$$

the equation becomes

$$HP = \frac{P}{6600} \frac{Q}{.2597},$$

and then

$$HP_f = \frac{PQ}{1714}.$$

This is the **fluid horsepower** formula where

$$HP_f = \text{Fluid } HP$$
$$P = \text{Pressure in psi}$$
$$Q = \text{Flow rate in GPM.}$$

Example: What fluid horsepower is needed in a system to deliver 10 GPM at 500 psi?

$$HP_f = \frac{PQ}{1714}$$

$$HP_f = \frac{(500)(10)}{1714}$$

$$HP_f = \frac{5000}{1714}$$

$$HP_f = 2.92 \; HP \; \text{(Ans.)}.$$

Since the pressure drop in the system is determined by an actual **measurement**, losses due to friction are already taken into consideration.

HEAT GENERATION

As energy is released, converted, or transmitted during system operation, part of it is converted into an unuseable form, primarily **heat**. Wiring in an electric motor, moving parts in a pump, system components, and conductors carrying fluid all generate heat which represents wasted energy. While a limited amount of heating may be needed to bring oil to best operating temperature (120°–140°F), most systems generate more than they can use. In larger systems, it is often necessary to provide cooling by means of heat exchangers. Excessive heat turns oil to sludge and damages system components and seals.

An estimate of **energy loss** in a system, or in an individual component, can be made using the **pressure drop** and the **flow rate** in a system while operating **without doing work**. With no load on the actuators, operate the pump and measure the pressure at the **pump discharge port**. Assuming the tank is not pressurized, this pressure reading will be the total pressure drop in the system. Using the rated pump output, the system energy loss is calculated using the **fluid horsepower** formula just derived:

$$HP_f = \frac{PQ}{1714}$$

where

HP_f = Energy loss, HP
P = System pressure drop
Q = Flow Rate, GPM.

We learned previously that **2545 BTUs/hr.** is equal to **one horse-**

power. Therefore, multiply this answer by **2545** to find the amount of thermal (heat) energy lost.

Example: How much heat energy is developed in a system delivering 20 GPM if the no-load pressure drop is 300 psi?

$$HP_f = \frac{PQ}{1714}$$

$$HP_f = \frac{(300)(20)}{1714}$$

$$HP_f = \frac{6000}{1714}$$

$$HP_f = 3.50 \; HP$$

$$3.50(2545) = 8907.5 \text{ BTU/hr. (Ans.).}$$

Heat generated in the system is radiated into the air surrounding the pump, piping, valves, actuators, and tank. This tends to keep temperatures in the oil and system components at acceptable levels. When necessary, cooling fans, fins, or heat exchangers are used to assist cooling.

Heat is generated whenever fluid flow is accompanied by a drop in pressure.

Major sources of heating are the **pressure relief** and **pressure control** valves, as pressure drops tend to be large across these components.

Remember, a positive-displacement pump must be constantly **moving fluid** whenever it is operating, and if this fluid is not being taken by actuators or other components, it must be directed back to the tank. Most applications have rest periods when the actuators are idle. During **short** idle periods, the fluid may be returned to the tank through the pressure relief valve. The calculation of energy loss due to flow through the valve begins with the same formula as above, but heat is generated **only** during the time the **actuators** are **idle.** The final step in the calculation is to multiply the figure for BTU/hr. by the **percentage** of time the actuators are **not doing work.**

Example: How much heat energy is produced by fluid flow through a pressure relief valve if the flow rate is 2.5 GPM, the pressure drop is 1200 psi, and the actuators operate 70% of the time?

$$HP_f = \frac{PQ}{1714}$$

$$HP_f = \frac{(1200)(2.5)}{1714}$$

$$HP_f = \frac{3000}{1714}$$

$$HP_f = 1.75 \; HP$$

$$1.75(2545) = 4454.5 \; \text{BTU/hr.}$$

This is the amount of heat lost if the fluid were flowing through the valve 100% of the time. In this problem, the actuators are working 70% of the time, so the flow through the valve takes place only 30%.

$$4454.5(.30) = 1336.4 \; \text{BTU/hr. (Ans.)}.$$

This is the average amount of heat produced by flow through that valve alone during the time of system operation. If the duty cycle of the system is such that flow through a pressure relief generates too much heat or lowers the system efficiency to an unacceptable level, an **unloading valve** may be used. This generates much less heat. Unloading valves will be explained in Chapter 7.

HEAT DISSIPATION

For hydraulic systems operating in very cold regions, some heating is desirable to bring the oil to best operating temperature. Sometimes heaters are used, especially when starting up. In most systems, however, too much heat is generated and it must be released. When heat is released from the system, it is said to be **dissipated**.

Ideally, **natural radiation** from the surfaces of the pump, piping, valves, actuators, tank, etc. is enough to keep the temperature at a reasonable level. Several factors contribute to overheating. The amount of heat **generated** may be too high. The **air temperature** at the work location may be high, **limiting** the amount of heat radiation. The installation may be near a furnace or melter, causing heat to be **added** to the system from outside sources. In these cases, some means in addition to natural radiation must be adopted to control fluid temperatures.

Under usual conditions, in which hydraulic equipment operates where people work in comfort and no significant heat is added

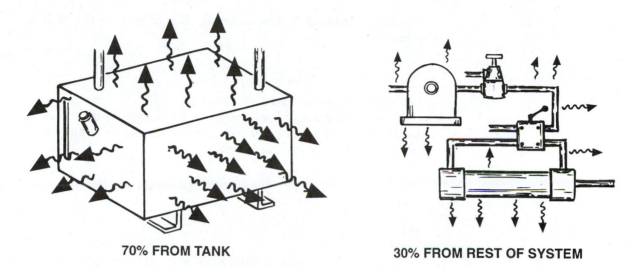

70% FROM TANK

30% FROM REST OF SYSTEM

from outside sources, about **70%** of the heat released is radiated from the surface of the **tank,** the other **30%** from the **piping** and **components.** Fortunately, hydraulic fluid carries heat from the components in which it is generated to the tank where it is most readily dissipated. As shown in Figure 5-6, the tank plays a major role in maintaining fluid temperature. The rate at which heat is dissipated through radiation from a surface to still air depends upon three factors:

Figure 5-6 Heat dissipation through radiation

1. the total **area** of the surface,
2. the **temperature difference** between the surface and the surrounding air, and
3. the surface **material.**

A large surface area radiates more heat than a small one. The amount of energy dissipated can be increased by using a **larger tank** or by mounting thin **fins** on its surface, similar to those on an air-cooled engine or air compressor. Installing a tank in a **cool room,** or **away** from **furnaces** or **heat-producing machinery** increases the temperature difference factor and helps cooling. A related solution is to direct air from a **blower** or **fan** onto the tank surface. This can remove as much as **twice** the heat energy as still air.

The material from which the tank is made, and its condition, are also factors in the amount of heat radiated. Rusted cast iron radiates more heat than new. Iron radiates more than copper. Through experiments in laboratories, figures called **constants** for various materials have been obtained, and we use these constants in formulas. The letter k represents a value called a **constant** which has been determined through laboratory testing. For most commercially produced hydraulic tanks, it has been found that a value of

$$k = .4$$

yields reasonably accurate solutions.

Given the amount of heat energy to be dissipated by the tank itself and the temperature of the surrounding air, we find the surface area needed using the formula

$$A = \frac{(k)(E)}{T_o - T_a}$$

where

A = Tank Surface Area in sq. ft.
k = Constant for the surface material
E = Heat Energy, BTU/hr.
T_o = Desired Oil Temperature, °F
T_a = Surrounding Air Temperature, °F.

Example: You have a hydraulic system which generates 4000 BTU/hr. The surrounding air temperature is 70°F, the maximum desired oil temperature is 140°F, and the tank is expected to remove 70% of the heat energy. First determine the amount of energy to be dissipated by the tank.

$$.70(4000 \text{ BTU/hr.}) = 2800 \text{ BTU/hr.}$$

$$A = \frac{(k)(E)}{T_o - T_a}$$

$$A = \frac{(.4)(2800)}{140 - 70}$$

$$A = \frac{1120}{70}$$

$$A = 16 \text{ sq. ft. (Ans.).}$$

You will need a tank with a total surface area, (top, bottom, and 4 sides) of at least 16 sq. ft., or 2304 sq. in.

FLUID VELOCITY

So far, fluid motion has been described in terms of its **flow rate**, measured in **gallons per minute**. There is another significant measurement with which we must be familiar, called **fluid velocity**.

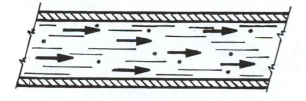

LAMINAR FLOW

TURBULENT FLOW

Figure 5-7 Two kinds of fluid flow

Velocity is defined as the distance traveled by an object in a specified time, or the time rate of motion in a stated direction.

Fluid velocity is usually represented as the distance in feet that a single particle, or drop, of fluid travels in one second. That is,

Velocity = ft./sec., or fps.

For a given flow rate (GPM), fluid travels at a higher velocity in a small pipe than in a large one, just as a river flows faster where it is shallow.

Fluid flow through piping generates heat by friction. Large pipe has more surface area and cools fluid more. Large pipe also causes less pressure drop as fluid flows through it. However, as a practical matter, designers specify pipe as small as possible to reduce cost and simplify installation. Usually the factor which determines the smallest acceptable size is the **fluid velocity**.

Normally, fluid flows through pipe in a relatively smooth, steady stream. This is called **laminar** flow. As the velocity is increased, it reaches an upper limit at which time flow becomes **turbulent**, much like river rapids, and efficiency drops significantly. This upper limit depends upon the density and viscosity of the fluid and size of the pipe. Engineers use these factors in a formula to calculate a figure called a **Reynolds Number** to predict whether flow in a given system will be laminar or turbulent. Characteristics of the two kinds of flow are illustrated in Figure 5-7.

Fluid velocity in a given line or pipe depends upon the **flow rate** in gallons per minute and the cross-sectional **area** of the pipe. One gallon equals 231 cubic inches. If one gallon of fluid were poured into a vertical pipe having a cross-sectional area of 1 square inch, it would make a vertical fluid column 231 inches high. If it were being poured at 1 GPM, the top of the column would be rising at a rate of 231 inches per **minute**.

Fluid **velocity** is measured in feet per **second**, or **fps**. A flow rate of 1 GPM is equal to $\frac{1}{60}$ gallon per second. Dividing by 60 in the example just given, the top of the column would rise at a rate of

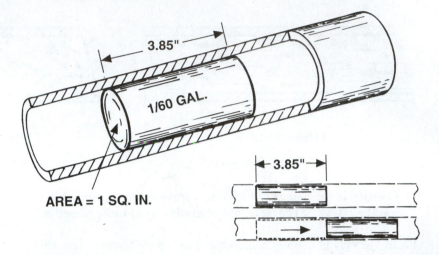

Figure 5-8 Displacement of $\frac{1}{60}$ gallon per second in a pipe

3.85"

1/60 GAL.

AREA = 1 SQ. IN.

3.85"

3.85 inches per second. For convenience in calculation, consider the movement of $\frac{1}{60}$ gallon of oil inside a cylinder having a cross-sectional area of 1 square inch. This much oil will be 3.85″ long as shown in Figure 5-8.

At a **flow rate** of **1 GPM**, this cylinder of oil moves its own length, or **3.85″**, in **one second.** It therefore has a **fluid velocity** of **3.85 in./sec.,** or **.3208 ft./sec.** If the cross-sectional area of the pipe were only **.5 sq. in.,** the $\frac{1}{60}$ gallon of oil would be a cylinder **twice** as long, or **7.70″.** That same flow rate of 1 GPM would now result in a fluid velocity of **7.70 in./sec.,** or **.6416 ft./sec.**

Doubling the flow rate to **2 GPM** in the pipe having an area of **1 sq. in.** would also result in a flow velocity of **.6416 ft. per second. Flow velocity** therefore depends upon the **flow rate** of the fluid and the **cross-sectional area** of the line, or pipe, and is calculated using the formula

$$V = \frac{.3208Q}{A}$$

where

V = Fluid Velocity, fps
Q = Flow Rate, GPM
A = Cross-sectional Area of the pipe.

Given the **diameter** of a pipe, we would find the cross-sectional area using the formula:

$$A = .785d^2.$$

Substituting this in the formula and simplifying, you get

$$V = \frac{.3208Q}{.785d^2} = \frac{.409Q}{d^2}$$

where

$$V = \text{Fluid Velocity, fps}$$
$$Q = \text{Flow Rate, GPM}$$
$$d = \text{inside diameter.}$$

Example: What is the fluid velocity of oil flowing at 3 GPM in a pipe having an inside diameter of .25"?

$$V = \frac{.409Q}{d^2}$$

$$V = \frac{.409(3)}{(.25)^2}$$

$$V = \frac{1.227}{.0625}$$

$$V = 19.63 \text{ fps (Ans.).}$$

Oil is caused to flow from a tank to a pump through a **pump inlet line** by **atmospheric pressure**. This pressure is about **14.7 psi** at sea level; less at higher elevations. The pump only **lowers** the pressure in the inlet line. It does not create a **perfect vacuum**, which would be zero pressure. It is the pressure **difference** between the tank and the pump that moves the oil. If the fluid velocity in the pump inlet line is too high, **air bubbles** may be drawn into the oil and damage the pump. This is called **cavitation** and will be explained in Chapter 6.

It has been determined experimentally that fluid velocity in the **pump inlet line** should be **2–5 fps**. For comparison, 5 fps is about as fast as a "brisk walk." To develop a formula for determining an appropriate inside diameter of a pump inlet line, begin with the fluid velocity formula:

$$V = \frac{.409Q}{d^2}$$

Since the velocity should not exceed **5 fps**, substitute that figure in the formula:

$$5 = \frac{.409Q}{d^2}$$

Multiplying both sides of the equation by d^2 and dividing by 5, you get

$$d^2 = \frac{.409Q}{5}$$

$$d^2 = .082\ Q$$

$$d = \sqrt{.082\ Q}$$

where

$d =$ inside diameter
$Q =$ Flow Rate, GPM.

Example: For a pump which delivers 6 GPM, what inside diameter of pipe in the pump inlet line would result in a flow velocity of 5 fps?

$$d = \sqrt{.082\ Q}$$

$$d = \sqrt{.082(6)}$$

$$d = \sqrt{.492}$$

$$d = .70''\ \text{diameter (Ans.).}$$

Fluid flow in the pump inlet line is caused by the relatively low pressure of the **atmosphere.** Flow in the pump discharge line is caused by the force of the **pump,** and may be subject to very high pressure. Here the problem is not air bubbles, but possible **turbulent flow.** The recommended upper limit for fluid velocity in a pump **discharge** line, including the system piping, is **15 fps** for operating pressures up to **500 psi** and **20 fps** for pressures between **500** and **2000 psi.** For comparison, the average velocity of an object dropped from a height of 15 ft. is about 15 fps. We will use this figure in our calculations. The formula for finding the smallest acceptable diameter for **system piping** then, is

$$d = \sqrt{.027\ Q}$$

where

$d =$ inside diameter
$Q =$ Flow Rate, GPM.

Note that all piping following the **discharge port** of the pump can be **smaller** than the pump **inlet line,** because the fluid velocity limit is **higher.**

Energy can neither be **created** nor **destroyed,** but the operation of nearly any working system requires that it be **released** or **converted** into a specific form. **Energy** is the **ability** to do **work. Mechanical** energy is measured in **horsepower. Electrical** energy is measured in **watts. Thermal,** or **heat** energy is measured in **BTU/hr.**

One horsepower is the amount of energy required to lift **550 lbs. 1 foot** in **1 second.**

$$HP = \frac{(F)(d)}{550t}$$

One horsepower equals 746 watts equals 2545 BTU/hr.

$$HP = \frac{W}{746}$$

$$HP = \frac{E}{2545}$$

$$HP = .00039\ E$$

Watts equals volts times amps.

$$W = VA$$

Efficiency is the percentage of input energy that is useful for a specific purpose compared with the **total** energy input.

$$\text{Efficiency} = \frac{\text{Output}}{\text{Input}} \times 100$$

$$\text{Output} = (\text{Efficiency})(\text{Input})$$

$$\text{Input} = \frac{\text{Output}}{\text{Efficiency}}$$

The **overall efficiency** of several components in a **series** is calculated by **multiplying** the efficiencies of all the components.

Cylinder horsepower is the amount of energy needed for a cylinder to function in a specific application.

$$HP_{cyl} = \frac{(F)(S)}{6600t}$$

Multiply the **calculated** cylinder horsepower by 1.25 to provide an allowance for friction losses when selecting a prime mover (motor) for an actual system. **Fluid horsepower** is the amount of energy **actually needed** to operate an **entire system,** and takes into account losses due to friction. The formula is based upon **actual** fluid flow and pressure drop.

$$HP_f = \frac{PQ}{1714}$$

This same formula is used to calculate energy **losses** in an operating system. Across a single component, such as a pressure relief valve, **measure** the **flow** through the valve and the **pressure drop.** For an entire system, measure the pressure P at the pump **discharge port** when the actuators are **not doing work.** For Q, use the rated pump output. This is the **no-load** energy used in the system, or **waste.**

Heat is generated whenever **fluid flow** is accompanied by a **pressure drop.** To determine **heat** energy loss, multiply the **no-load** horsepower figure by 2545:

$$HP(2545) = \text{BTU/hr., or } E.$$

Most systems generate more heat than they can use, and normally this is **dissipated** through **radiation.** The rate of heat dissipation depends on (1) the total surface **area,** (2) the **temperature difference,** and (3) the surface **material.** About **70%** of the heat dissipated is radiated from the tank; **30%** is radiated from the piping and other components.

To determine the surface area needed to dissipate a given amount of heat energy, use the formula

$$A = \frac{(k)(E)}{T_o - T_a}$$

where

A = surface area in sq. ft.
k = a "constant" for the surface material
E = heat energy to be dissipated in BTU/hr.
T_o = oil temperature in °F
T_a = surrounding air temperature in °F.

For most commercially available tanks, the figure for the constant is

$$k = .4.$$

When necessary, additional cooling can be accomplished by **fins**, a **fan**, or a **heat exchanger**. For most systems, the recommended fluid temperature is **120°–140°F**.

Fluid velocity is the distance a single particle of fluid travels in one second.

$$V = \text{ft./sec., or fps.}$$

Fluid velocity in a line, or pipe, depends upon the **flow rate** and the **cross-sectional area** of the line, and is calculated using the formula

$$V = \frac{.3208Q}{A}$$

where

$V =$ Flow Velocity in fps
$Q =$ Flow Rate, GPM
$A =$ Cross-sectional Area in sq. in.

Movement of fluid through a pipe in a smooth, steady stream, is called **laminar flow**. When the fluid velocity is too high, the flow becomes agitated, or "stirred up" and is called **turbulent**. Turbulent flow wastes energy and generates heat. **Atmospheric pressure** moves fluid from the tank through the pump inlet line when **suction** is created in the pump. Fluid velocity in the **pump inlet line** should be **2–5 fps**, to avoid drawing **air bubbles** into the fluid. To find an appropriate **pump inlet line** diameter use the formula

$$d = \sqrt{.082Q}.$$

Fluid velocity in the **discharge** side of the pump, including the high-pressure working lines of the system, should be about **15 fps**, to avoid **turbulent flow**. To find an appropriate diameter for all piping on the **discharge** side of the pump, use the formula

$$d = \sqrt{.027Q}.$$

PROBLEMS

[handwritten: HP = ω/746 = I ∝ O · O/I × 100 = Eff]

5.1 What is the efficiency of an electric motor that uses 829 watts of energy to produce 1.0 HP?

5.2 What is the efficiency of an electric motor that uses 1570 watts of energy to produce 2.0 HP?

5.3 What is the efficiency of an electric motor that uses 2027 watts of energy to produce 2.5 HP?

5.4 What is the efficiency of an electric motor that uses 1467 watts of energy to produce 1.75 HP? *[handwritten: VA = ω   HP = ω/746   O/I × 100 = Eff]*

5.5 What is the efficiency of a 220 volt electric motor that draws 4.0 amps and produces 1.0 HP?

5.6 What is the efficiency of a 220 volt electric motor that draws 6.2 amps and produces 1.75 HP?

5.7 What is the efficiency of a 440 volt electric motor that draws 3.6 amps and produces 2.0 HP?

5.8 What is the efficiency of a 440 volt electric motor that draws 2.8 amps and produces 1.5 HP?

5.9 You have a hydraulic system in which an electric motor drives a pump. The motor *[handwritten: (Em)(Ep) = Ec]* operates at 95% efficiency while the pump operates at 82% efficiency. What is the efficiency of the combination of motor and pump?

5.10 You have a hydraulic system in which an electric motor drives a pump. The motor operates at 92% efficiency while the pump operates at 80% efficiency. What is the efficiency of the combination of motor and pump?

5.11 You have a hydraulic system in which an electric motor drives a pump. The motor operates at 96% efficiency while the pump operates at 78% efficiency. What is the efficiency of the combination of motor and pump?

5.12 You have a hydraulic system in which an electric motor drives a pump. The motor operates at 94% efficiency while the pump operates at 85% efficiency. What is the efficiency of the combination of motor and pump?

5.13 You have a hydraulic system in which the prime mover is a 220 volt electric motor which draws 6.3 amps and operates at 95% efficiency. What is the HP output of this motor? *[handwritten: VA = ω ω/746 · Eff = O/I]*

5.14 You have a hydraulic system in which the prime mover is a 220 volt electric motor which draws 7.5 amps and operates at 92% efficiency. What is the HP output of this motor?

5.15 You have a hydraulic system in which the prime mover is a 440 volt electric motor which draws 1.75 amps and operates at 90% efficiency. What is the HP output of this motor?

5.16 You have a hydraulic system in which the prime mover is a 440 volt electric motor which draws 7.6 amps and operates at 88% efficiency. What is the HP output of this motor?

   For questions 17–20, assume that available motors range from .5 HP to 8.0 HP in increments of .5 HP (.5 HP, 1.0 HP, 1.5 HP, 2.0 HP, 2.5 HP, etc.). They all operate at 90% efficiency.

5.17 You are assembling a system whose actuators (cylinders) require 2.3 hydraulic HP to operate. If the efficiency of the pump is 82%, what size motor should be used? Add 25% to your calculated HP figure to allow for friction losses.

$HP = \dfrac{O}{Eff} \times 1.25$

overall first

5.18 You are assembling a system whose actuators (cylinders) require 4.0 hydraulic HP to operate. If the efficiency of the pump is 78%, what size motor should be used? Add 25% to your calculated HP figure to allow for friction losses.

5.19 You are assembling a system whose actuators (cylinders) require 1.8 hydraulic HP to operate. If the efficiency of the pump is 85%, what size motor should be used? Add 25% to your calculated HP figure to allow for friction losses.

5.20 You are assembling a system whose actuators (cylinders) require 5.0 hydraulic HP to operate. If the efficiency of the pump is 76%, what size motor should be used? Add 25% to your calculated HP figure to allow for friction losses.

5.21 How many HP are required to lift a 7000 lb. load 40 ft. in 5 minutes?

$HP = \dfrac{(F)(d)}{t}$ *lbs. ft sec*

5.22 How many HP are required to lift a 4500 lb. load 25 ft. in 2 minutes?

5.23 How many HP are required to lift a 5000 lb. load 30 ft. in 30 seconds?

5.24 How many HP are required to lift a 180 lb. load 150 ft. in 90 seconds?

5.25 You have a hydraulic circuit in which a piston rod must extend 10 inches in 3 seconds with a force of 1500 lbs. How many HP will this require?

$HP_{cyl} = \dfrac{FS}{6600t}$

5.26 You have a hydraulic circuit in which a piston rod must extend 14 inches in 1.2 seconds with a force of 900 lbs. How many HP will this require?

5.27 You have a hydraulic circuit in which a piston rod must extend 18 inches in 1.5 seconds with a force of 700 lbs. How many HP will this require?

5.28 You have a hydraulic circuit in which a piston rod must extend 15 inches in .8 seconds with a force of 500 lbs. How many HP will this require?

5.29 You have a 3″ diameter, 10″ stroke cylinder installed in a system with a hydraulic pressure of 500 psi.

$V = (A)(S)$ $HP = \dfrac{PV}{6600t}$

   (a) How much force can be exerted when the rod extends?
   (b) How many HP are required to move a load 8 inches in 1.5 seconds with this much force?

5.30 You have a 4.5″ diameter, 12″ stroke cylinder installed in a system with a hydraulic pressure of 600 psi.
   (a) How much force can be exerted when the rod extends?
   (b) How many HP are required to move a load 9 inches in 2.2 seconds with this much force?

$F = PA$

5.31 You have a 2.5″ diameter, 8″ stroke cylinder installed in a system with a hydraulic pressure of 850 psi.
(a) How much force can be exerted when the rod extends?
(b) How many HP are required to move a load 5 inches in 1.5 seconds with this much force?

5.32 You have a 5″ diameter, 24″ stroke cylinder installed in a system with a hydraulic pressure of 600 psi.
(a) How much force can be exerted when the rod extends?
(b) How many HP are required to move a load 12 inches in 2 seconds with this much force?

5.33 What fluid HP is needed to deliver 8.5 GPM at 600 psi?

$$HP = \frac{PQ}{1714}$$

5.34 What fluid HP is needed to deliver 4.0 GPM at 1000 psi?

5.35 What fluid HP is needed to deliver 7.5 GPM at 1800 psi?

5.36 What fluid HP is needed to deliver 2.4 GPM at 2000 psi?

5.37 How much heat energy is developed in a system delivering 4 GPM if the no-load pressure drop is 50 psi? $HP = \frac{PQ}{1714} \times 2545$

5.38 How much heat energy is developed in a system delivering 7 GPM if the no-load pressure drop is 85 psi?

5.39 How much heat energy is developed in a system delivering 5.5 GPM if the no-load pressure drop is 92 psi?

5.40 How much heat energy is developed in a system delivering 6.0 GPM if the no-load pressure drop is 73 psi?

5.41 How much heat energy is produced by fluid flow through a pressure relief valve if the flow rate is 3.0 GPM, the pressure drop is 1500 psi, and the actuators operate 65% of the time? $100 - 65 = R$ $HP = \frac{PQ}{1714} \times 2545 \times$

5.42 How much heat energy is produced by fluid flow through a pressure relief valve if the flow rate is 4.5 GPM, the pressure drop is 1000 psi, and the actuators operate 38% of the time?

5.43 How much heat energy is produced by fluid flow through a pressure relief valve if the flow rate is 6.5 GPM, the pressure drop is 800 psi, and the actuators operate 25% of the time?

5.44 How much heat energy is produced by fluid flow through a pressure relief valve if the flow rate is 2.5 GPM, the pressure drop is 420 psi, and the actuators operate 30% of the time?

5.45 You have a hydraulic pump which delivers 5 GPM.
(a) What inside diameter pipe in the low pressure inlet line will result in a flow velocity of 5 ft. per second?
(b) What inside diameter of pipe in the high pressure discharge line will result in a flow velocity of 15 ft. per second?

5.46 You have a hydraulic pump which delivers 3.5 GPM.
    (a) What inside diameter pipe in the low pressure inlet line will result in a flow velocity of 5 ft. per second?
    (b) What inside diameter of pipe in the high pressure discharge line will result in a flow velocity of 15 ft. per second?

5.47 You have a hydraulic pump which delivers .5 GPM.
    (a) What inside diameter pipe in the low pressure inlet line will result in a flow velocity of 5 ft. per second?
    (b) What inside diameter of pipe in the high pressure discharge line will result in a flow velocity of 15 ft. per second?

5.48 You have a hydraulic pump which delivers 9.0 GPM.
    (a) What inside diameter pipe in the low pressure inlet line will result in a flow velocity of 5 ft. per second? $d = \sqrt{.082 Q}$
    (b) What inside diameter of pipe in the high pressure discharge line will result in a flow velocity of 15 ft. per second? $d = \sqrt{.027 Q}$

Pumps

6

The operation of any hydraulic system is based upon the ability of a **moving fluid** to transmit **energy** to **do work**. The device that **causes** the fluid to move is called a **pump**.

> **The purpose of a hydraulic pump is to cause fluid to flow. It does not, by itself, create pressure. Pressure results only when there is resistance to flow.**

This is an important concept in the study of hydraulics. As Figure 6-1 illustrates, you have probably put this to practical use with a garden hose without realizing it. It can be readily verified by inserting a pressure gauge in an open-ended pump discharge line. Any pressure reading will be the result of friction in the line following the gauge.

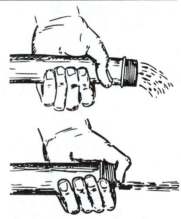

Figure 6-1 Resistance to flow causes pressure

PUMP CLASSIFICATION

Nearly all pumps used in industrial hydraulics are classified as **positive-displacement pumps**. This means that a **specific amount** of fluid is displaced with each revolution or stroke of the pump, and it **must** be allowed to leave through the discharge port. If blocked, either the prime mover (motor or engine) will **stall** or something must **break**. Fluid not used is normally directed back to the tank.

Non-positive-displacement pumps move an **indefinite** amount of fluid with each revolution, and if flow is blocked, the fluid is merely agitated, much like water in a washing machine. These are suitable for low pressure, high volume applications, where their

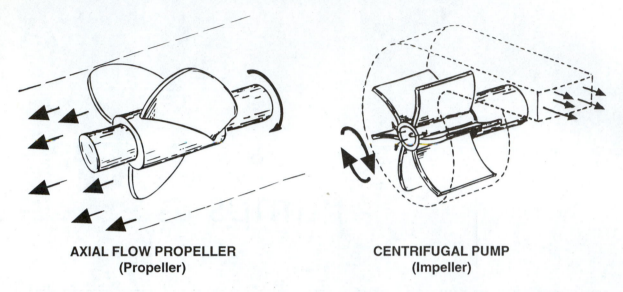

AXIAL FLOW PROPELLER
(Propeller)

CENTRIFUGAL PUMP
(Impeller)

Figure 6-2 Non-positive-displacement pumps

function is to **move fluid** rather than do work. Typical applications include pumps in home hot water heating systems, coolant pumps on machine tools, and fresh water pumps on ships. Examples are **centrifugal** pumps, which direct fluid **outward** from the shaft, and **axial** pumps, which direct fluid in a path **parallel** to the shaft. These are both shown in Figure 6-2.

POSITIVE-DISPLACEMENT PUMPS

Three basic types of **positive-displacement pumps** are used in hydraulic systems: (1) **gear** pumps, (2) **vane** pumps, and (3) **piston** pumps. While the three are very different in **appearance**, the **operation** of each consists of a common three-step sequence:

1. A relative **vacuum** is created within the pump, allowing **atmospheric pressure** to move fluid from the reservoir into the **inlet port** of the pump;
2. Fluid within the pump is confined within a totally enclosed space, or **chamber,** and isolated from the inlet port. It is this isolation from the inlet port that allows high pressure to develop in the **discharge** line; and
3. Fluid is forced out of the discharge port.

The rate at which fluid flows from the pump depends upon the **size** of each chamber and the **rate** at which the individual chambers are being "emptied" through the discharge port. In a gear pump like the one in Figure 6-3, for example, the spaces between the gear teeth are "chambers," and the RPM of the gears determines the flow.

Figure 6-3 External gear pump. *Courtesy of Danfoss Fluid Power, Racine, WI*

The amount of fluid moved by a pump in one revolution, or stroke, is called its volumetric displacement, or simply displacement.

VOLUMETRIC DISPLACEMENT

In some pumps, the displacement can be varied, either manually or automatically in response to system pressure. Pumps whose displacement is not adjustable are called **fixed-displacement** pumps. Those which can be changed are called **variable-displacement** pumps. Those variable displacement pumps which can be **automatically** adjusted by system pressure are called **pressure-compensated** pumps. In any pump, some fluid **leaks back** within the pump because of **system pressure**. In a gear pump, this reverse flow occurs between the teeth of the two gears as they come together, between the teeth and the housing, and along the sides of the gears. This internal leakage, called **slippage**, is illustrated in Figure 6-4.

Pump manufacturers assign **speed** and **pressure ratings** to their products, and there is little slippage when operating within the normal operating range. When the specified limits are exceeded, slippage increases rapidly, heat is generated, and pump output is

Figure 6-4 Slippage in an external gear pump

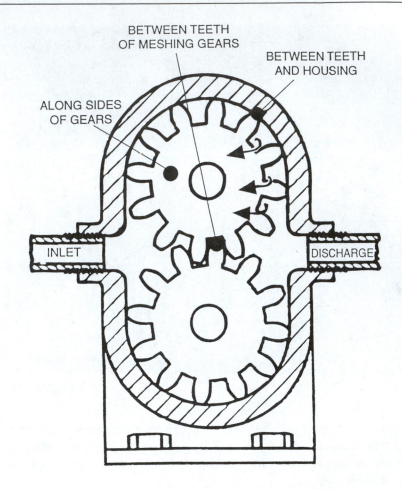

BETWEEN TEETH
OF MESHING GEARS

BETWEEN TEETH
AND HOUSING

ALONG SIDES
OF GEARS

INLET

DISCHARGE

Figure 6-5 Effect of pressure on pump output

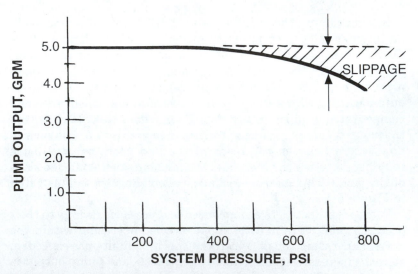

reduced. The decrease in output flow with increased pressure is shown in Figure 6-5. Pumps sometimes have **dual ratings.** One is for **continuous** operation and one is for **intermittent** (interrupted) operation which allows time for cooling.

VOLUMETRIC EFFICIENCY

A limited amount of slippage is desirable, as it lubricates moving parts. Since it does represent a decrease in pump output, however, energy is wasted. A comparison between **actual** pump output and **ideal** pump output (zero slippage) is called **volumetric efficiency.**

Volumetric efficiency is the percentage of actual pump output compared with what the ouput would be if there were no slippage.

Since slippage results from system pressure, if there were **no** load (resistance to flow) on the pump there would be no **slippage.** Actually, there is always some resistance, due to friction in the discharge port.

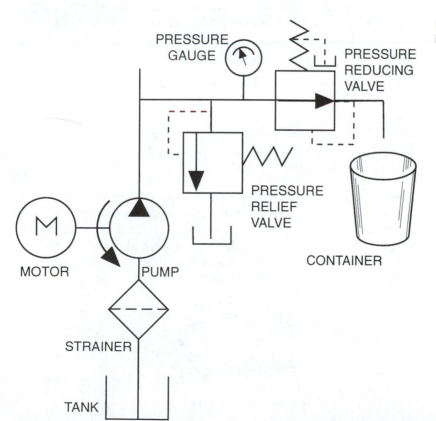

Figure 6-6 Circuit for determining volumetric efficiency

PRESSURE GAUGE

PRESSURE REDUCING VALVE

PRESSURE RELIEF VALVE

CONTAINER

MOTOR

PUMP

STRAINER

TANK

A reasonably accurate calculation of volumetric efficiency of a given pump can be made using **no-load** and **rated-pressure** GPM outputs using this procedure:

1. Assemble a circuit which includes a **pressure relief** valve followed by a **pressure reducing** valve. Attach a short **line** to the discharge port of the pressure reducing valve. Set the **pressure relief** valve to a pressure **higher** than the pump rating.
2. With the pressure reducing valve open **all the way**, start the pump and let it run a few minutes.
3. Check to see that **no** oil is passing through the **pressure relief** valve, then fill a container from the line and **record** the **time**. The pump is now delivering oil at **no-load. Hint:** Using as **large** a container as is convenient will increase the accuracy of your calculation.
4. Now adjust the **pressure reducing** valve to the **rated pressure** of the pump. If necessary, **readjust** the **pressure relief** valve so that **no** oil passes through it back to the tank.
5. Again fill the container from the line and **record** the **time**. The pump is now delivering oil at **rated pressure**.
6. **Calculate** the **volumetric efficiency** using the formula

$$\text{Efficiency} = \frac{T_1}{T_2}$$

where

T_1 = Time to fill a container at zero pressure
T_2 = Time to fill the container from the pressure relief valve at rated pressure

Example: You have a pump rated at 1.0 GPM at 1500 psi. With the pressure reducing valve fully open, and the pump delivering oil at no load, you fill a 3 gallon container in 3 minutes and 3 seconds. With the pressure reducing valve set at the pump rating, you fill the container in 3 minutes and 30 seconds. What is the volumetric efficiency of this pump?

$$\text{Efficiency} = \frac{T_1}{T_2}$$

$$\text{Efficiency} = \frac{183 \text{ sec.}}{210 \text{ sec.}}$$

$$\text{Efficiency} = 87\% \text{ (Ans.).}$$

As compared with **variable-displacement** pumps, **fixed-displacement** pumps generally have **higher** volumetric efficiencies and are **quieter** because of their more **rigid construction**. The various **movable** parts which **make** a pump output adjustable provide **more** leakage paths inside a variable-displacement pump. In any system, the **pump rating** should match the **GPM required** for system operation as closely as possible. When **more** oil is pumped than the system uses, **energy** is wasted and the **heat** generated can be harmful to the oil.

CAVITATION

Under certain conditions of **heat, lowered pressure,** or **turbulence,** gas or air **bubbles** called **cavities** are created in fluids. We see such cavities in carbonated soft drinks, on the bottom of a pan of boiling water, and on the surface of boat propellers. This process is called **cavitation** and can be very damaging to mechanical devices such as ship propellers and hydraulic pumps.

Cavitation is a process in which bubbles of air or gas are created in fluid as a result of heat, lowered pressure, or turbulence (see Figure 6-7).

Extensive research on cavitation has been conducted, primarily by those involved in the design of propellers for large ships. Still

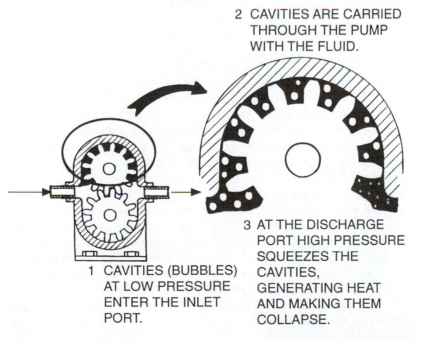

2 CAVITIES ARE CARRIED THROUGH THE PUMP WITH THE FLUID.

1 CAVITIES (BUBBLES) AT LOW PRESSURE ENTER THE INLET PORT.

3 AT THE DISCHARGE PORT HIGH PRESSURE SQUEEZES THE CAVITIES, GENERATING HEAT AND MAKING THEM COLLAPSE.

Figure 6-7 Cavitation

the process is not fully understood. For over 70 years, it was generally believed that the erosion damage from cavitation took place **only** when the bubbles **collapsed**, as a result of fluid impacting surfaces. More recent studies indicate that damage **may** occur at the point where cavities are **formed**. There is even evidence that some corrosion results from **electrolytic** action. This is the process that rusts cars in regions where roads are salted in winter. Whatever the exact mechanism, engineers have long been aware of the damaging effects of cavitation and offer these **recommendations:**

1. **Check the oil level** in the reservoir **regularly,** adding when necessary to ensure that no air is drawn into the pump inlet line;
2. Use a pump **inlet line** having a **diameter large enough** so the inlet flow velocity never exceeds **5 ft./sec.;**
3. Keep the pump **inlet line short** and with as **few fittings** as possible;
4. Place the **filter** in the reservoir **return line** rather than the pump inlet, if possible;
5. If the **filter** must be in the pump inlet line, choose a type with **low pressure-drop** and an **indicator** showing when it becomes dirty and needs to be replaced; and
6. Keep the **oil temperature** below **150°F.**

Many kinds of pumps have been developed and used for a wide variety of purposes. It has been said that more patents have been issued for pumps than for any other single class of device. In this chapter we will examine only a few of the most common to learn their principles of operation and characteristics.

GEAR PUMPS

The **external gear pump,** previously discussed in Chapter 4, normally consists of meshing gears within a housing. **Most pumps** use ordinary **spur gears.** On others, **helical** or **herringbone** teeth are used for **quieter operation** and **improved performance** but at **greater expense.** The **gear pump** is by far the **most popular** of the three basic types because it is **ruggedly constructed,** provides **reasonable efficiency** (80%–90%), is tolerant of **dirty oil,** and is generally the **least expensive.**

Very similar in operation to the external gear pump is the **lobe pump** as shown in Figure 6-8, which resembles an external gear pump with only three teeth on each gear. Both shafts must be connected to the prime mover (motor) however, since the lobes do not mesh like gears. **Lobe pumps** are generally **quieter** than gear pumps and have a greater **volumetric displacement,** but the **flow** is **not as smooth.** The flowing fluid surges, or throbs, just as blood pulsates when a heart beats.

Figure 6-8 Lobe pump

INLET

DISCHARGE

Figure 6-9 Internal gear pump

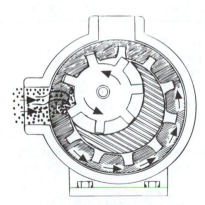

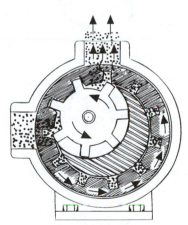

SEPARATING TEETH
CREATE A PARTIAL
VACUUM, ALLOWING
ATMOSPHERIC
PRESSURE TO FORCE
FLUID INTO THE PUMP.

FLUID IS CARRIED
THROUGH THE PUMP IN
CHAMBERS BETWEEN
THE TEETH OF THE
INTERNAL GEAR.

AS TEETH OF THE
EXTERNAL GEAR MESH
WITH THE TEETH OF
THE INTERNAL GEAR,
FLUID IS FORCED OUT.

Another type is the **internal gear pump** shown in Figure 6-9, in which a small **external gear** rotates within a larger, **internal gear**. Suction is created at the inlet port as the gear teeth separate, leaving vacant spaces, or chambers, between the teeth of the **internal gear**. As the gears continue to rotate, the fluid is held in the chambers by a crescent, or half-moon shaped divider. The gear rotates, carrying the fluid through the pump to the discharge port. There the teeth of the smaller (external) gear again fill the chambers, leaving no room for the fluid, and it is forced out the discharge port.

All gear and lobe pumps are **fixed-displacement** pumps, which means that the **amount** of **fluid** moved in **one revolution cannot be changed.** The output at a given speed (RPM) is always the same. **Some vane** and **piston** pumps are constructed so the output can be varied **without** changing the **speed (RPM).** When made this way, they are called **variable-displacement** vane or piston pumps.

VANE PUMPS

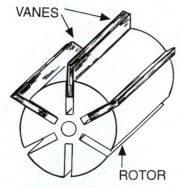

VANES

ROTOR

Figure 6-10 Vane pump rotor and vanes

Figure 6-11 Unbalanced fixed-displacement vane pump

A type of pump with long lasting moving parts and most suitable for applications requiring a relatively **high volume** of fluid at **medium pressure** is the **vane pump. Vane pumps** generally have **greater volumetric displacement** than gear pumps, but run at **slower speeds.** With proper **maintenance,** the vanes do not **wear** as quickly as gears, but the **fluid** must be kept **clean,** and **effective lubrication** is more **critical.** They are available as either **fixed-displacement** or **variable-displacement,** and some have a **pressure-compensation** feature that allows the output to be changed automatically to meet system demand.

As Figure 6-10 illustrates, the **moving** part of the vane pump consists of a **rotor** driven by a shaft, with slots holding **sliding vanes.** As the rotor spins, the vanes slide outward in the slots and press against the inner surface of the pump housing, forming chambers for carrying fluid. In some pumps, **pressurized oil** is directed into the bottom of the slots to help centrifugal force in pushing the vanes outward.

The simplest type is the **unbalanced, fixed-displacement** vane pump shown in Figure 6-11, having only one inlet and one outlet

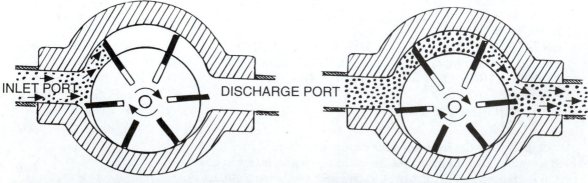

INLET PORT

DISCHARGE PORT

PARTIAL VACUUM AT THE INLET PORT ALLOWS ATMOSPHERIC PRESSURE TO FORCE FLUID IN.

FORCE EXERTED BY THE PRIME MOVER ON THE VANES MOVES FLUID FROM THE DISCHARGE PORT.

port. It is the **least expensive** of the vane pumps and suitable for **lower** system **pressures**. The high-pressure fluid is concentrated on one side of the drive shaft, however, resulting in an **unbalanced** force on the shaft bearings. This causes friction and wear.

A **movable ring** is placed within the pump housing to create a **variable-displacement vane pump.** An **adjustment screw** is used to **move** the **ring** with respect to the rotor. This changes the **size** of the **chambers** between vanes, and therefore the **amount of fluid** moved with each revolution of the rotor. With this adjustment, the **volumetric displacement** can be made to vary from the **maximum** pump output rating down to **zero** (see Figure 6-12).

In a similar type vane pump, a small **piston** inside a cylinder replaces the **adjustment screw.** A **pilot line** from the pump discharge port feeds **fluid** back to **actuate** the piston. When system **pressure rises** high enough to compress the spring, it moves the ring **toward the center** and **reduces** the pump **output.** On this type, an adjustment screw on the spring is used to set system pressure. Normally no pressure relief valve is needed, thereby **saving energy** and **reducing** the likelihood of **oil breakdown** from excessive heat. This feature is called **pressure-compensation,** and pumps so equipped are called **pressure-compensated vane pumps.**

As you can see in Figure 6-13, a **balanced, fixed-displacement vane pump** has **two inlets** and **two discharge ports.** The high pressure fluid becomes concentrated in **two** areas, on **opposite** sides of

Figure 6-12 Unbalanced variable-displacement vane pump

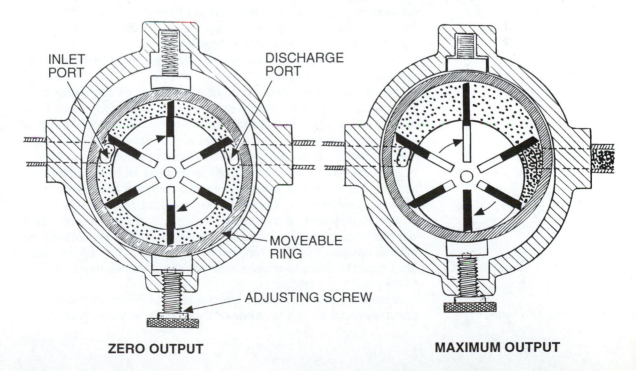

INLET PORT

DISCHARGE PORT

MOVEABLE RING

ADJUSTING SCREW

ZERO OUTPUT

MAXIMUM OUTPUT

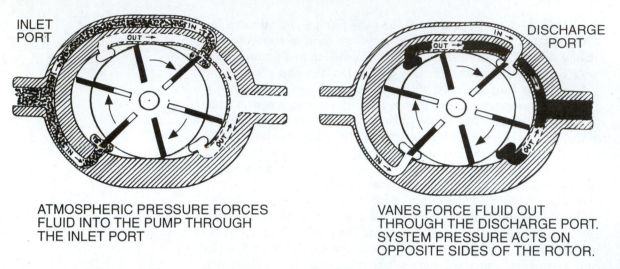

INLET PORT

DISCHARGE PORT

ATMOSPHERIC PRESSURE FORCES FLUID INTO THE PUMP THROUGH THE INLET PORT

VANES FORCE FLUID OUT THROUGH THE DISCHARGE PORT. SYSTEM PRESSURE ACTS ON OPPOSITE SIDES OF THE ROTOR.

Figure 6-13 Balanced fixed-displacement vane pump

the drive shaft. The forces are now **balanced** so there is no side loading of the shaft. This enables the pump to handle relatively high pressure, but the **volumetric displacement** is **constant**. That is, the GPM output of a **balanced** vane pump cannot be changed except by changing the speed (RPM).

PISTON PUMPS

According to historical records, the first devices developed to do **work** with fluid were early forms of piston pumps. Bramah's jack, described in Chapter 1, was actually a **reciprocating** piston pump with a **single** check valve. The modern **bottle jack**, available in hardware stores, is called a **lever-operated, single-acting, reciprocating piston pump.**

In general, the various types of **piston pumps** are **superior** in performance to either the gear or vane pumps just described. They are capable of operating at speeds as high as **10,000 RPM** and of supplying systems having pressures up to about **15,000 psi**. **Volumetric efficiencies** range up to 98%. They are generally very durable and operate for extended periods without failure. The nature of their construction, however, is more **complex** than with other types and requires very **close tolerances**. This makes their initial **cost** relatively **high**. Furthermore, when they do break down they normally must be returned to the **manufacturer** for repair because of their complexity. This, too, can be expensive. Their applications are limited to those in which the need for high efficiency and reliability justify the higher costs.

Several kinds of piston pumps are used in hydraulic power applications, but all fall into one of three **general** classifications:

1. **Reciprocating plunger** piston pumps,
2. **Radial** piston pumps,
3. **Axial** piston pumps.

Radial and axial piston pumps are **rotary** pumps because on each the drive shaft **rotates**, or **turns.** They differ in the direction in which their pistons, or plungers, move with respect to the drive shaft. In a **radial** pump, their motion is on a line extending **outward**, or "radially" from the centerline of the shaft. In an **axial** pump, movement is **parallel**, or "**axial**" to the shaft centerline.

RECIPROCATING PLUNGER PISTON PUMPS

This simplest type consists of a piston, or plunger, inside a cylinder with inlet and discharge lines. A check valve is located in each line. As the piston rod is extended and retracted, fluid is alternately drawn into the cylinder through the inlet line and then forced out through the discharge line. Its operation is similar to that of a hand-operated bicycle tire pump. This is called a **single-acting** pump and is depicted in Figure 6-14. In a variation of this, called a **double-acting** pump, inlet and discharge lines are attached to both the cap end and rod end, causing fluid to be drawn in and discharged during both extension and retraction. A double-acting pump is illustrated in Figure 6-15.

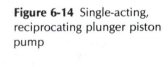

Figure 6-14 Single-acting, reciprocating plunger piston pump

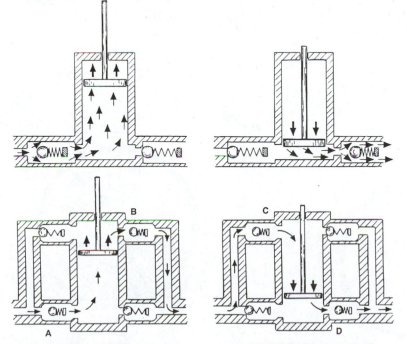

Figure 6-15 Double-acting, reciprocating plunger piston pump

FLUID ENTERS THROUGH *A* AND IS PUMPED OUT THROUGH *B.*

FLUID ENTERS THROUGH *C* AND IS PUMPED OUT THROUGH *D.*

RADIAL PISTON PUMPS

In some respects, the motion within a radial piston pump is like that of a vane pump. One popular design uses what is called a **pintle valve,** consisting of inlet and outlet ports **centered** in the **housing.** A **cylinder block,** similar to a vane pump **rotor,** rotates on a pintle, or pin, within a **reaction ring** whose center is **offset** from the **pintle.** This block has either **seven** or **nine** holes, or "bores" whose center-lines extend outward from the shaft. **Sliding plungers** move within the holes in the cylinder block as they ride against the inner surface of the reaction ring. The pumping action takes place **inside** the **cylinder block** however, rather than in **chambers** as is the case with the **vane pump.** On **variable-displacement pumps,** the amount of **offset** and therefore the **GPM output** can be changed either **manually,** or automatically by **system pressure.** When accomplished automatically, the pump is called a **pressure-compensated** radial piston pump.

As the offset is reduced to the point where the cylinder block and reaction ring have the **same centers,** the plungers stop moving in and out, and the pump output becomes **zero.** When a **pressure-compensated** pump is used, there is normally **no need** for a **pressure relief** valve. The inlet and discharge lines are ported at the **center** of the pump, and the plungers **alternately** draw fluid **in** from the **inlet** and force it **out** through the **discharge ports** as the rotor turns. Figure 6-16 shows the construction of a radial piston pump and the means by which the output is changed.

Figure 6-16 Radial piston pump

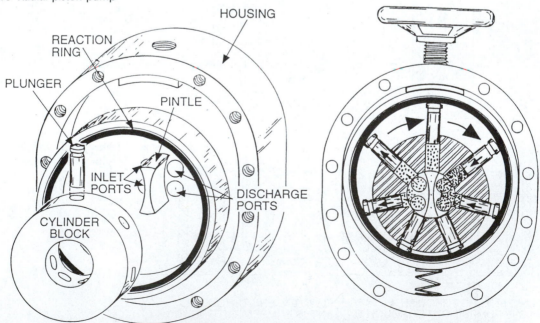

In an axial piston pump, the movement of the pistons, and therefore the flow of the fluid, is **parallel** to the axis, or shaft. Either **seven** or **nine** pistons inside a **cylinder block** revolve around a shaft much like bullets in the cylinder of a revolver, or "six-gun" seen in western movies. One design of an axial piston pump is shown in Figure 6-17.

Pump manufacturers provide a wide variety of axial pumps, with both fixed and variable displacement capability. They are classified as **in-line** or **bent-axis** types, depending upon the way the pistons are made to reciprocate within their bores. An **in-line** pump makes use of a **cam** or **angled plate** upon which the ends of the pistons slide. A **bent-axis** pump uses two shafts connected with a **universal joint,** similar to that used on the drive shafts of cars and trucks. Both types are available as either fixed or variable displacement pumps. The difference between in-line and bent-axis pumps is illustrated in Figure 6-18.

In the **swash plate** pump, an example of the **in-line** type, the revolving pistons are moved in and out by a disk called a **swash plate** mounted at an angle to the axis. In some pumps, the swash plate is mounted permanently in place. This type has a **fixed-displacement.** In others, it is mounted in a **yoke** allowing the angle to be changed. The displacement is therefore **variable,** and the GPM output can be set at any amount between zero and the maximum pump rating.

AXIAL PISTON PUMPS

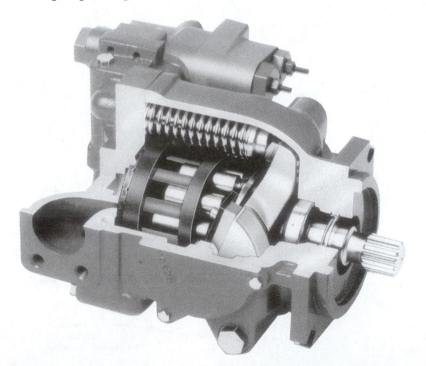

Figure 6-17 Axial piston pump construction. *Courtesy of Vickers Inc., Troy, MI*

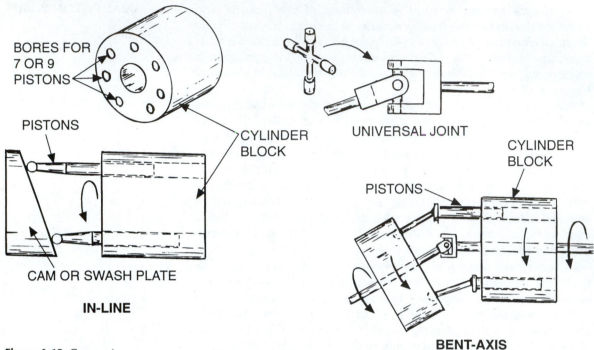

BORES FOR
7 OR 9
PISTONS

PISTONS

CYLINDER
BLOCK

UNIVERSAL JOINT

CYLINDER
BLOCK

PISTONS

CAM OR SWASH PLATE

IN-LINE

BENT-AXIS

Figure 6-18 Comparison between in-line and bent-axis construction

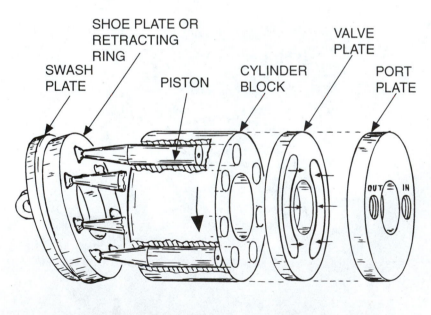

SHOE PLATE OR
RETRACTING
RING

SWASH
PLATE

PISTON

CYLINDER
BLOCK

VALVE
PLATE

PORT
PLATE

OUT   IN

Figure 6-19 Variable-displacement, swash plate type, axial piston pump

WEEP HOLE FOR
LUBRICATION

As the **cylinder block** revolves, the **shoe plate** rotates with it, sliding on the **swash plate**. The **pistons**, attached to the **shoe plate**, "reciprocate," or slide in and out of the holes in the block. Their movement draws fluid from the tank into the **inlet port** during one half revolution, and forces it out the **discharge port** during the other half. To reduce friction between the shoe plate and the swash plate, a small hole called a **weep hole** in each piston carries oil to the sliding surfaces. The **tilt angle** between the swash plate and the drive shaft centerline determines the **volumetric displacement** of the pump. Flow control by swashplate adjustment is shown in Figure 6-20. Both in-line and bent-axis pumps are available with pressure compensation, normally eliminating the need for a pressure relief valve.

Schematic symbols do **not** indicate whether the pump used is a gear, vane, or piston. As you can see in Figure 6-21, the symbol on a drawing indicates whether the pump is **fixed-displacement**, **variable-displacement**, or **pressure-compensated**. Other data such as the rating, manufacturer, and model will be included in the installation specifications.

Figure 6-20 (Below) Swash plate tilt determines the pump displacement

Figure 6-21 (Bottom) Pump symbols

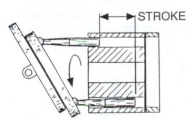

STEEPEST ANGLE
LONGEST STROKE
HIGHEST GPM OUTPUT

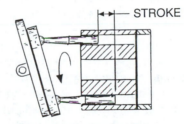

REDUCED ANGLE,
SHORTENED STROKE
LESS GPM OUTPUT

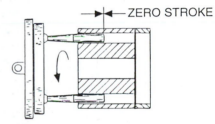

ZERO ANGLE
NO MOVEMENT
NO OUTPUT

FIXED DISPLACEMENT

**VARIABLE DISPLACEMENT
NON-COMPENSATED**

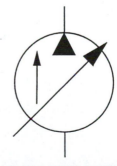

**VARIABLE DISPLACEMENT
PRESSURE-COMPENSATED**

CHAPTER SUMMARY

The **purpose** of a **hydraulic pump** is to cause **fluid to flow**. It does **not**, by itself, create **pressure**. **Pressure** is the result of **resistance** to flow. Nearly all pumps used in industrial hydraulics are **positive-displacement** pumps. **Non-positive-displacement** pumps, on the other hand, are suitable for low-pressure, high-volume applications where their function is to **move fluid** rather than do work. The three basic types of **positive-displacement** pumps are (1) **gear** pumps, (2) **vane** pumps, and (3) **piston** pumps, and the operation of them involves a **three-step sequence:**

1. A **vacuum** is created internally, allowing **atmospheric pressure** to force fluid into the **inlet port;**
2. Once inside the pump, the fluid is **confined** in a totally enclosed space and **isolated** from the inlet port;
3. Moving elements inside the pump **force** the fluid out the **discharge port.**

The amount of fluid moved by a pump in **one revolution** or **one stroke** is called its **volumetric displacement,** or simply its **displacement.** Pumps are classified as **fixed-displacement** or **variable-displacement,** depending upon whether the output per revolution or stroke can be changed. Internal **reverse flow** of flow of fluid, as a result of system pressure, is called **slippage.** A **limited amount** of slippage is desirable, to provide lubrication. The percentage of **actual** pump displacement as compared with what the displacement **would be** if there were **no slippage** is called **volumetric efficiency.**

Volumetric efficiency is calculated by taking flow measurements, then **dividing** the **time** it takes to fill a given container at **no load** with the time it takes at **rated pump pressure. Cavitation** is a destructive process in which bubbles, or cavities, of air or gas are created due to **heat, lowered pressure,** or **turbulence.** The operating procedures recommended to avoid **pump damage** due to **cavitation** are:

1. Check the tank **oil level** regularly;
2. Use an **inlet line** large enough so the inlet **flow velocity** does not exceed **5 ft./sec.;**
3. Use a **short** inlet line, with few fittings;
4. Install the **filter** in the pump **return** line;
5. Use a **low pressure-drop** filter with an **indicator** showing when it needs replacing, if it has to be installed in the pump **inlet** line;
6. Keep the oil **temperature** below **150°F.**

The **gear** pump is the **most popular** because of its rugged **construction,** reasonable **efficiency, tolerance** for dirty oil, and low **cost. Lobe** pumps are similar to gear pumps, but the lobes do not mesh like gear teeth, and both shafts must be driven. As compared with gear pumps, they are generally **quieter**, they have a greater **volumetric displacement**, and the flow **pulsates**, or **surges** more. All **gear** and **lobe** pumps are **fixed-displacement** types.

Vane pumps are generally more durable because the moving parts are less subject to wear. They are most suitable for applications involving high fluid **volume** at medium **pressure**. They tend to have greater **volumetric displacement** than gear pumps, but run at slower **speeds**. With vane pumps, **clean oil** and **effective lubrication** are more critical. Vane pumps are available as either **fixed** or **variable** displacement units. Some have **pressure-compensation**, which means they vary their output automatically to meet system demand. A vane pump with only one inlet and one discharge port is **unbalanced**, because system pressure on the discharge side puts force against the drive shaft causing friction and wear. A **balanced** vane pump has **two** inlet and **two** discharge ports, causing **equal** and **opposite** forces on the shaft. Only **fixed-displacement** pumps can be **balanced**.

Of the three basic types, **piston pumps** provide the best performance. They can operate at the highest **speeds** and **pressure**, have the highest **volumetric efficiencies**, and operate for extended periods without failure. Offsetting these advantages are their high **cost**, both in purchase price and maintenance. There are **three** general classifications of **piston** pumps: (1) **reciprocating plunger**, (2) **radial**, and (3) **axial**.

Reciprocating plunger piston pumps can be either **single-** or **double-acting**. **Single-acting** pumps deliver fluid only when the rod retracts. **Double-acting** pumps deliver fluid during both extension and retraction. In **radial** piston pumps, either **seven** or **nine** pistons reciprocate **radially** within a cylinder block whose center is offset from the center of either the **housing**, or a **reaction ring**. As the block rotates, the ends of the pistons are pushed in and out of their bores, drawing fluid in the **inlet** port and forcing it out the **discharge** port of the pump. Like vane pumps, **radial** pumps are available with both **fixed** and **variable-displacement** capability and with **pressure-compensation**. In **axial** piston pumps, either **seven** or **nine** pistons move **parallel** to the drive shaft inside a **cylinder block**. **In-line** axial piston pumps use either an **inclined cam** or a **tilted disk** called a **swash plate** to generate piston movement inside the **cylinder block**. The swash plate **tilt angle** may be made adjustable, providing **variable-displacement** capability. If this adjustment is automatic, to respond to system demand, the pump is said to be **pressure-compensated**.

A **universal joint** in the **bent-axis** type axial piston pump provides the pumping action. **Variable-displacement** capability is achieved by changing the angle at which the two shafts intersect. When this is accomplished automatically, the pump is said to be pressure-compensated. **Schematic symbols** on a drawing indicate only whether a pump is **fixed-displacement**, **variable-displacement**, or **pressure-compensated**. Specific information such as whether it is gear, vane or piston, the speed, capacity and pressure ratings, make and model, and manufacturer will be found in the installation specifications.

PROBLEMS

6.1 What is the purpose of a hydraulic pump?

6.2 What causes pressure in a hydraulic system?

6.3 Name the three basic types of positive-displacement pumps most commonly used in industrial applications.

6.4 Why are non-positive-displacement pumps not usually appropriate for industrial hydraulic applications?

6.5 Explain briefly the difference between positive-displacement and non-positive-displacement pumps.

6.6 Describe the three-step sequence that takes place during the operation of a positive-displacement pump.

6.7 Explain briefly the difference between radial and axial motion.

6.8 What causes fluid to flow from the tank to the inlet port of a positive-displacement pump?

6.9 What is the term applied to reverse flow inside a positive-displacement pump resulting from system pressure? Why is a limited amount of this desirable?

6.10 Explain what is meant by "volumetric displacement."

6.11 Explain what is meant by "volumetric efficiency."

6.12 Why is it important for the GPM output of a pump to closely match the needs of the system if operation is to be continuous?

6.13 How would you determine the volumetric efficiency of a pump in a laboratory?

6.14 Explain briefly what is meant by "cavitation."

6.15 Explain why cavitation is undesirable, and describe three ways of preventing it.

6.16 Explain, with a sketch, how an external gear pump works.

6.17 Explain the difference between a fixed-displacement pump and a variable-displacement pump.

6.18 List three reasons why gear pumps are the most commonly used of the three basic types.

6.19 Give two reasons why you might use a lobe pump instead of a gear pump for a given application.

6.20 Which of the three basic types of positive-displacement hydraulic pumps is not available as a variable-displacement pump?

6.21 List two advantages and two disadvantages of using a variable-displacement pump in a system.

6.22 What is the function of pressure compensation in a hydraulic pump?

6.23 Why must a pressure relief valve be provided in a hydraulic system having a non-compensated, fixed-displacement pump?

6.24 Explain briefly why it is desirable to have a vane pump balanced, and how this is accomplished.

6.25 Identify which of the three basic hydraulic pumps is most efficient and give two reasons why it isn't used for all applications.

6.26 Make a sketch of a double-acting, reciprocating piston pump and explain how it works.

6.27 Name the three basic types of piston pumps.

6.28 You have a pump rated at 1.0 GPM at 1000 psi. It takes 2 minutes and 2 seconds to fill a 2 gallon container at no load. At rated system pressure, it takes 2 minutes and 19 seconds to fill the same container. What is its volumetric efficiency?

6.29 You have a pump rated at 1.5 GPM at 1800 psi. It takes 1 minute and 21 seconds to fill a 2 gallon container at no load. At rated system pressure, it takes 1 minute and 29 seconds to fill the same container. What is its volumetric efficiency?

6.30 You have a pump rated at .75 GPM at 500 psi. It takes 2 minutes and 24 seconds to fill a 2 gallon container at no load. At rated system pressure, it takes 2 minutes and 55 seconds to fill the same container. What is its volumetric efficiency?

6.31 You have a pump rated at 3.0 GPM at 500 psi. It takes 1 minute and 45 seconds to fill a 5 gallon container at no load. At rated system pressure, it takes 2 minutes to fill the same container. What is its volumetric efficiency?

6.32 How is the surface of a swash plate in an axial piston pump lubricated?

6.33 How many pistons are used in a swash plate (in-line) type of axial piston pump?

6.34 Explain the meaning of the terms, "single-acting" and "double-acting" as they apply to piston pumps.

6.35 What kind of valve can be omitted from a system with a variable-displacement pump? Explain why.

Valving

7

If the **fluid** in a hydraulic system is to perform the **functions** for which the circuit is designed, it must be made to go **where** needed, **when** needed, and with an appropriate **pressure** and **flow rate** to do the job. This is the function of **valving**. Valves serve three specific functions:

1. They control **flow** so that fluid is delivered in the **amount** needed, **when** needed;
2. They control **pressure** to **protect** the circuit and personnel, and to limit the **force** exerted by **actuators**;
3. They **direct flow** to **where** the fluid is needed.

Quite logically then, valves are grouped according to function and categorized as **flow control, pressure control,** and **directional control** valves. As we shall see, however, some serve **more** than one purpose, and their capabilities, when used in **combination**, seem limited only by the **imagination** of system **designers**. Valves are manufactured in a variety of **forms**, each having **characteristics** to meet specific **needs**. In this chapter we will learn how a **few** of the **most common** valves work, and **why** certain types are preferred over others for specific **applications**.

VALVE OPERATION

Whenever the **position** of a valve is **changed**, the valve is said to be **operated**, or **actuated**. Valve **positions** include (1) **open** position, (2) **closed** position, and (3) any set of **flow paths**. In turning on a faucet, we are **operating** it by changing from the **closed** position to **open**. When a pressure relief valve **opens** as a result of high system pressure, it is said to be **actuated**.

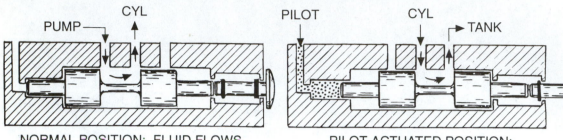

NORMAL POSITION: FLUID FLOWS
FROM THE PUMP TO THE CYLINDER.

PILOT-ACTUATED POSITION:
FLUID FLOWS FROM THE
CYLINDER TO THE TANK.
MANUAL OPERATION OF THE
PUSHBUTTON RETURNS THE
VALVE TO THE NORMAL POSITION.

Figure 7-1 Pilot-operated, push-button-return, directional control valve

The position a valve takes when at rest, or when the system is not in operation, is called its **normal** position. Valves can, for example, be **normally open, normally closed,** or **normally centered.** Sometimes a valve has no normal position, but simply remains in the position to which last set.

There are several ways to change a valve position, but all are categorized as either (1) **manual,** (2) **mechanical,** (3) **pilot,** or (4) **electrical.**

Manually operated valves have handles, knobs, or pushbuttons turned or moved by hand. Some valves have a **manual override.** This means that although they are normally operated by **other** means, they can be operated by hand when necessary. **Mechani-**

Figure 7-2 Operating principle of a solenoid valve

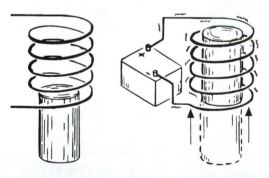

ELECTRIC CURRENT FLOWING THROUGH
A WIRE CREATES A MAGNETIC FIELD.
ANYTHING MADE OF IRON IS PULLED
INTO IT, AND HELD THERE, UNTIL
CURRENT STOPS FLOWING.

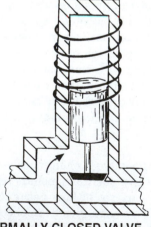

NORMALLY CLOSED VALVE
OPENS WHEN CURRENT
FLOWS.

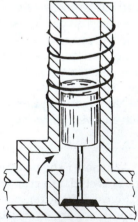

NORMALLY OPEN VALVE
CLOSES WHEN CURRENT
FLOWS.

Figure 7-3 Solenoid-operated directional control valves. *Courtesy of Vickers, Inc., Troy, MI*

cally operated valves are similar in construction to manual valves, except that a **mechanical device** such as a cam or trip lever changes the valve position rather than a **person**. **Pilot operated** valves, like the ones in Figure 7-1, have lines or passages through which **pressurized fluid** (oil or air) flows to move a valve **spool**. **Air-piloting** has replaced **electrical** operation in many applications to avoid potential **shock** and **fire** hazards, **maintenance** costs, and problems of compliance with **safety regulations**. Most **electrically operated** valves use a **solenoid** to move a valve element **magnetically** and change its position.

> The purpose of a flow control valve is to regulate the rate of fluid flow in a specific part of a circuit.

FLOW CONTROL VALVES

The usual purpose of this function is to govern the motion of the system's actuators, such as the pistons in hydraulic cylinders.

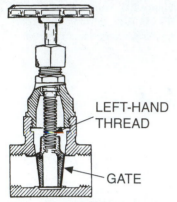

LEFT-HAND
THREAD

GATE

CLOSED
LEFT HAND THREAD
CAUSES THE GATE
TO CLOSE WHEN
THE HANDLE IS
TURNED CLOCKWISE.
THE TAPERED GATE
FITS TIGHTLY INTO
A TAPERED VALVE
SEAT.

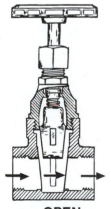

OPEN
WHEN FULLY OPEN,
FLUID FLOWS IN A
STRAIGHT PATH
WITH LITTLE
PRESSURE DROP.

Figure 7-4 Gate valve

By using **individual** flow control valves, **several** actuators operating at **different speeds** can be powered by a single hydraulic pump. The **position** of a flow control valve determines **whether** the actuator moves at all, as well as the **rate,** or **speed** of the movement.

The simplest flow control valve is the **gate valve** shown in Figure 7-4. It is intended to be used either **fully open** or **fully closed,** in applications where a part of a circuit sometimes needs to be isolated. When **open,** the straight-through flow path causes very little pressure drop. When **closed,** a wedge-shaped element, called a **gate,** fits into a seat to provide positive closure, blocking fluid flow. It is **not** meant to be used **partially open,** as the gate causes high turbulence and pressure drop and is hard to move. **Gate valves** are particularly suited for **high-pressure, high-volume** applications, and tolerate **high temperatures.** Their disadvantages are that they tend to **vibrate** when in a partially open position, **operation** is comparatively **slow,** both the gate and seat tend to **wear,** and the fluid used must be **clean** or leakage will result. Any small particle caught between the gate and its seat is likely to cause leakage.

Somewhat similar in appearance to the gate valve, but different in operation and purpose, is the **globe valve** shown in Figure 7-6. Here the sealing element is a **disk** or **cone,** which moves **against** the direction of flow when closing. Globe valves are suitable for **high pressure** applications, and are commonly used **partially open** for control from zero to full flow. This is called **throttling,** or **metering.** However, there is considerably more **pressure drop** than in a gate valve because the fluid must make **sharp turns** while passing through. **Flow** through a globe valve is intended to be in **one direction only.** An **arrow** on the side of the housing indicates the **direction** of flow. As with gate valves, **clean fluid** is critical in a system using **globe** valves because small particles can prevent proper seating and cause **leakage.**

Ball valves are adaptable to a wide variety of applications and are characterized by low **pressure** drop, low **leakage,** light **weight,** and small **size** for their flow and pressure ratings. Because they open and close with a "wiping" motion, they can **tolerate** more **contamination** from grit and particles than most other types. They can be designed to handle highly **viscous** ("thick") and **corrosive** fluids. Because actuation consists of only a 90° ball rotation, they can be **opened** and **closed** very quickly. Ball valve operation is illustrated in Figure 7-7. The **disadvantages** of ball valves are their high **cost,** the tendency of the valve seat **material** to be **squeezed** out of position when used for **throttling,** and sudden **pressure variations** (called "water hammer" by plumbers) from rapid opening or closing.

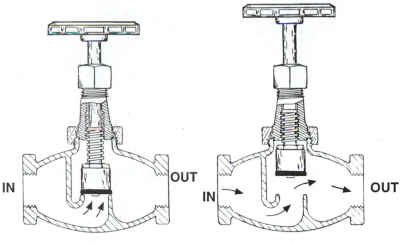

WHEN INSTALLED
CORRECTLY, FLUID FLOWS
AGAINST THE STEM.

FLOWING FLUID MUST
MAKE SHARP TURNS,
CAUSING TURBULENCE
AND BACK PRESSURE.

Figure 7-5 (Left) Handwheel-operated gate valve. *Courtesy of Kennedy Valve, Elmira, NY*

Figure 7-6 (Right) Globe valve

Figure 7-7 (Below) Ball valve

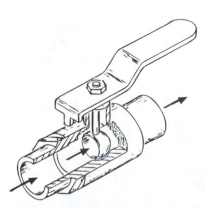

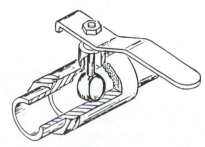

OPEN
FLUID FLOWS FREELY
THROUGH OPENING
IN THE BALL.

CLOSED
90° ROTATION OF
THE HANDLE CLOSES
THE VALVE.

Figure 7-8 Taper plug valve

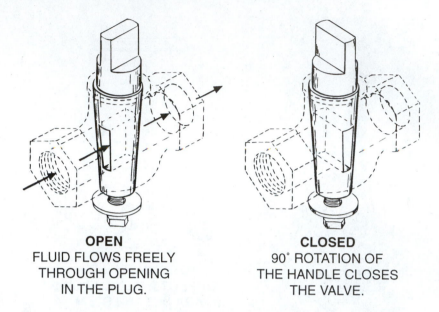

OPEN
FLUID FLOWS FREELY
THROUGH OPENING
IN THE PLUG.

CLOSED
90° ROTATION OF
THE HANDLE CLOSES
THE VALVE.

Taper plug valves are quite similar to ball valves except that their actuating element is a **tapered cylindrical plug** which rotates 90° to control fluid flow. Like ball valves, they are light in weight and small in size for their flow and pressure ratings. The **wiping action** as they open and close provides **tolerance** for grit and particles in the fluid. Since a change from full-open to full-close involves only a single 90° motion, **response is quick**. Finally, they are **cheaper** than ball valves because of their simpler construction (see Figure 7-8).

Because effective sealing is accomplished by a "wedging" action of the taper, taper plug valves have a tendency to bind. Some are **lubricated** to reduce this fault, but they must be **serviced** regularly and care must be taken to ensure that the lubricating material does not **contaminate** the fluid in the system.

When system operation requires **precise** flow control, with the flow **rate** subject to **change**, a **needle valve** is used. In this type, **throttling** or **metering** is accomplished by means of a narrow tapered stem called a **needle** positioned in line with a tapered **hole**, or **orifice**. Because of the two tapers, the change in the size of the opening is very gradual. This provides a **higher** degree of **control** than is possible with globe valves.

In many applications, it is desirable to have precise flow control in **one direction** only, with the return flow **unrestricted**. For example, the rod of a cylinder may need to advance a cutting tool at a slow, controlled rate and then return quickly in time for the next piece. Needle valves are available with built-in **check valves** to meet this requirement. Figure 7-9 depicts this feature. All fluid

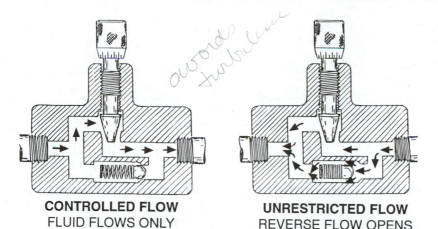

Figure 7-9 Needle valve with built-in check valve

CONTROLLED FLOW
FLUID FLOWS ONLY THROUGH THE NEEDLE-CONTROLLED PASSAGE.

UNRESTRICTED FLOW
REVERSE FLOW OPENS THE CHECK VALVE.

Figure 7-10 Needle valves. *Courtesy of Deltrol Fluid Products, Bellwood, IL*

flow in the **controlled** direction passes through the **needle valve** because flow in the bypass is **blocked** by the **check valve. Reverse** flow passes freely through the **check valve,** and **also** through the **needle valve** if it is not fully closed. An **arrow** on the side of the valve indicates the direction of **controlled flow.**

Since pressure is the result of **resistance** to fluid flow, **partially closing** any flow control valve creates a **pressure drop.** Furthermore, any **variation** in the **load** on an actuator **changes** the **amount**

Figure 7-11 Pressure-compensated flow control valve (controlled orifice)

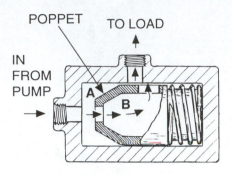

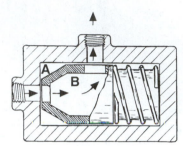

POPPET   TO LOAD

IN
FROM
PUMP

A   B

A   B

DURING NORMAL OPERATION
FLUID FLOW PUSHES POPPET
AGAINST THE SPRING
BECAUSE
THE PRESSURE AT **A** IS
GREATER THAN THE
PRESSURE AT **B**.

AN INCREASE IN THE LOAD
INCREASES THE PRESSURE
AT **B**, REDUCING THE
PRESSURE DIFFERENCE.
THE SPRING IS NOW ABLE
TO MOVE THE POPPET
AND ENLARGE THE
OPENING.

of that pressure drop, and therefore changes the **flow rate**. In all of the types of valves described so far, any **increase** in **load** results in a **decrease** in **flow rate**. This means that the actuator (such as the piston rod in a cylinder) will **slow down**. In applications where the load is **constant**, or where the variation **does not affect** system performance, this is not a problem. The flow control valve is **set** for the proper flow rate **with the known load**.

Some valves **automatically** maintain a **constant** flow rate despite load changes. These are called **pressure-compensated** valves. One type is called a **controlled-orifice valve**. Two holes (orifices), one in a moving element called a **poppet** and the other in the valve **housing** are normally held slightly **offset** from each other by a spring. Any increase in **pressure** as a result of increased **load** releases the spring, widening the passageway formed by the two holes. This **compensates** for the change in pressure, keeping the flow rate **constant**. In an alternate version of the type shown in Figure 7-11, a spring **adjustment** is provided so the flow rate can be **changed**.

PRESSURE CONTROL VALVES

The purpose of a pressure control valve is to regulate fluid pressure, either for the protection of the system and personnel, or for the proper operation of the actuators.

You may find it helpful to **review** some of the fluid power **principles** explained in previous chapters before trying to understand

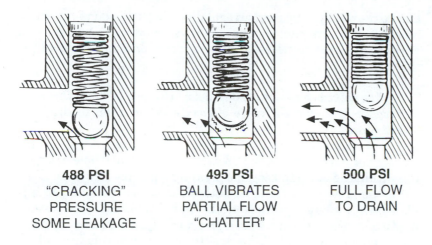

488 PSI
"CRACKING"
PRESSURE
SOME LEAKAGE

495 PSI
BALL VIBRATES
PARTIAL FLOW
"CHATTER"

500 PSI
FULL FLOW
TO DRAIN

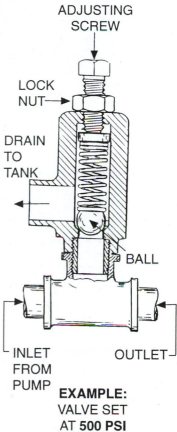

ADJUSTING
SCREW

LOCK
NUT→

DRAIN
TO
TANK

BALL

INLET
FROM
PUMP

OUTLET

EXAMPLE:
VALVE SET
AT **500 PSI**

Figure 7-12 Chatter in a simple pressure relief valve

pressure control. For example, the significance of **area** in determining how much **force** exerted under a given **pressure**

$$F = PA$$

is a critical factor in the operation of pressure control valves. You must remember that **pressure** is not developed at the **pump**, but at some point **downstream** if you are to understand how **pressure** can be regulated **without** changing pump **output**. Pressure control valves are classified as (1) **simple**, or **direct-acting** and (2) **compound**, or **piloted**.

In a **simple** valve, fluid pressure acts against a spring **poppet** and causes valve **actuation** at a set pressure. The pressure relief valve described in Chapter 4 is a **simple**, or **direct-acting** pressure relief valve. While suitable for most applications, it can cause problems wherever proper operation requires **precise** pressure control, or where **vibrations** in the flow cannot be tolerated. As pressure approaches the spring setting in a simple pressure relief valve, the ball tends to lift off its seat, allowing leakage. This is called the **cracking pressure**. Further increase in pressure causes the ball to "float" in the stream and **vibrate**, or **chatter**. This is shown in Figure 7-12. The pressure at which the valve becomes **fully open**, allowing **full** fluid flow back to the tank is called the valve **setting**. Therefore, a pressure **range** exists between the point where the valve **begins** to open and where it is **fully** open. **Within** this **range**, the ball **chatters** and flow is uneven.

When system operation requires **smooth** flow at a **precise** pressure, a **compound pressure relief valve** must be used. This includes some installations with **pressure-compensated** pumps. The valve reacts to change more **quickly** than the rather bulky pump mechanism, providing more **precise** regulation. In a compound pressure relief valve, overpressure opens a **pilot valve**, which in turn causes a **balanced piston** to release the pressure by draining the fluid to the tank.

A pilot valve is a device used to define the flow of fluid whose purpose is to control valve elements rather than to drive actuators.

During normal operation, fluid flows through a small hole in a piston within the valve to keep the pressure equal on both sides. Because the pressure is equal on both sides, it is called a **balanced piston**. Only a **light** spring is needed to hold this balanced piston in place. A **heavy** spring, with a **pressure adjusting screw** holds a **pilot valve** in place, blocking flow to the drain through another passage.

As Figure 7-13 illustrates, the system pressure rises and fluid flows through the small **hole** in the **balanced piston** to maintain equal pressure on both sides. When it exceeds the **set point** of the valve, as established by the **adjusting screw**, the **pilot valve** compresses the **heavy spring** and opens to allow fluid to flow from the chamber **above** the **balanced piston** to the **drain**. With the fluid released, pressure **drops** quickly, and the piston is **no longer** in **balance**. The hole in its center is **too small** to allow enough fluid flow to recover. **System pressure** now **lifts** the piston, which had only the **light spring** holding it in place, and opens the passage to the drain. The actuation is **quick**, with no "**range**" of **pressure** and no **vibration** as with a simple, or direct-acting pressure relief valve.

The obvious reason why a **compound** valve is not chosen for **all** installations is that these valves, being more **complex**, are more **expensive**. A second disadvantage is that **cleanliness** of the **fluid** in the system is more **critical**. A small particle of **dirt** or **grit** can prevent the valve from functioning, creating a **safety hazard**. A disadvantage shared by **both** types of pressure relief valves is that fluid at **full** system **pressure** flows through the valve passage to drain. This **wastes** energy which is lost as **heat**. This may not be enough to be significant in systems where the oil is **always**, or **nearly always** doing **work**. However, when the actual work cycle is short, oil is returned to the tank. The energy waste and potentially damaging heat then become **unacceptable**.

An **unloading valve** is often used in systems when a **substantial** amount of oil is drained during normal operation, such as those having **delays** between **cycles**. This is a type of **compound** valve which, when actuated, allows **free flow** with **very little** pressure drop. The **prime mover**, whether an electric motor or a gas or diesel engine, uses **less energy** when the pump moves free-flowing fluid.

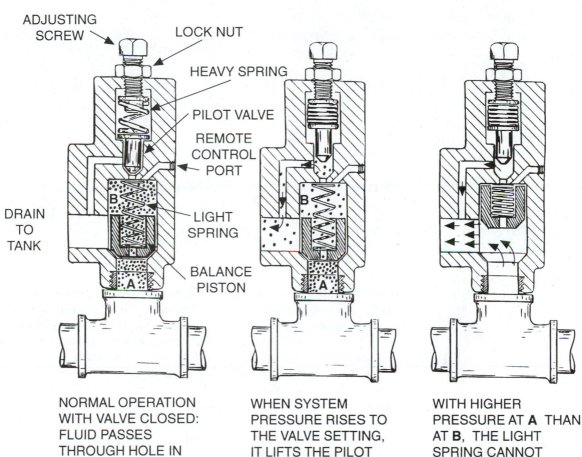

ADJUSTING
SCREW

LOCK NUT

HEAVY SPRING

PILOT VALVE

REMOTE
CONTROL
PORT

DRAIN
TO
TANK

LIGHT
SPRING

BALANCE
PISTON

B

A

NORMAL OPERATION
WITH VALVE CLOSED:
FLUID PASSES
THROUGH HOLE IN
BALANCED PISTON TO
MAINTAIN EQUAL
PRESSURE AT **A** AND **B**.

WHEN SYSTEM
PRESSURE RISES TO
THE VALVE SETTING,
IT LIFTS THE PILOT
VALVE, AND FLUID
DRAINS QUICKLY.
PRESSURE AT **B**
DROPS.

WITH HIGHER
PRESSURE AT **A** THAN
AT **B**, THE LIGHT
SPRING CANNOT
HOLD THE BALANCED
PISTON DOWN. FULL
FLOW NOW GOES TO
THE DRAIN.

Figure 7-13 Compound pressure relief valve

An electric motor draws less current, and an engine uses less gasoline or diesel fuel. In Figure 7-14, we see the sequence of operations by which oil is diverted to the tank through an unloading valve. During **normal** operation, oil flows into the valve, through a **check** valve, and out to the **load.** Like the compound relief valve, it has a **piston** balanced by **equal pressure** on top and bottom, and held in place by a **light spring.** Any pressure **change** takes place on **both** sides of the piston, **maintaining** this balanced condition.

Whenever system pressure exceeds the **set point** established by an **adjustment screw,** it lifts a **poppet** off its seat. This allows oil to drain from the **upper** side of the piston, through a small **hole** through its **center,** to the **tank.** As it drains, pressure on the **upper**

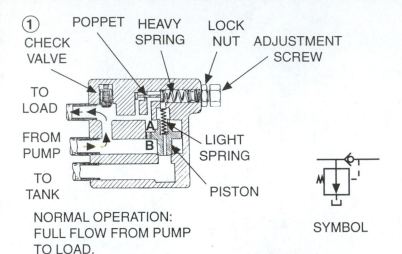

① CHECK VALVE — POPPET — HEAVY SPRING — LOCK NUT — ADJUSTMENT SCREW

TO LOAD

FROM PUMP

TO TANK

A

B

LIGHT SPRING

PISTON

NORMAL OPERATION: FULL FLOW FROM PUMP TO LOAD.

SYMBOL

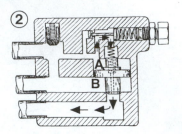

② A

B

SYSTEM PRESSURE RISES. POPPET COMPRESSES HEAVY SPRING. FLUID FLOWS FROM **A** THROUGH PISTON TO TANK.

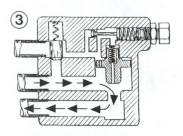

③

FLOW FROM **A** TO TANK CAUSES PRESSURE DROP. PISTON RISES, LETTING FLUID FLOW FROM PUMP TO TANK. CHECK VALVE BLOCKS RETURN FLOW FROM THE SYSTEM.

Figure 7-14 Unloading valve

side drops while pressure on the **lower** side remains high. The piston is now **unbalanced** and lifts off its seat. This puts the valve in the **unloading** state, and oil flows from the **pump** directly to the **tank.** There is very little **resistance** to flow, and therefore very little **pressure** in this passageway to keep the **check valve** open. The check valve closes, maintaining **system pressure** in the line to the **load.** The valve **continues** to unload oil directly to the tank until **system pressure drops,** normally as a result of the actuators taking oil from the line to do **work.** When system pressure drops to about **85%–90%** of the valve **setting,** the check valve reopens and normal operation **resumes.**

Sometimes a system is required to direct oil to **two** cylinders in a specific **order.** For proper operation, the **first** cylinder must complete its motion **before** the valve switches flow to the **second** cylinder. For example, a hydraulic **clamp** needs to grip a **workpiece** before a hydraulically driven **cutting tool** is advanced. A hydraulically operated **safety door** needs to be closed before a hydraulic **press** moves. For such applications a **sequence valve** may be used.

A **sequence valve** is a **compound** valve which **directs** flow through **one** passageway and **blocks** flow through another until **pressure** in a **pilot** line moves a **spool** (see Figure 7-15). A **remote control** port is included in the valve body so that the spool can be shifted by pressure from **another source** if necessary. This might be used, for example, in troubleshooting. A **check valve** may also be made part of the valve for convenience in returning the actuators to their original positions.

Some applications require that a valve be made to **lift** a load, then **block** return flow to **support** it in place hydraulically. For

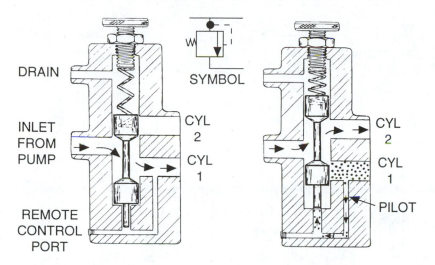

Figure 7-15 Sequence valve

DRAIN

SYMBOL

INLET FROM PUMP

CYL 2

CYL 1

REMOTE CONTROL PORT

PILOT

FLUID FLOWS FREELY THROUGH THE VALVE TO ACTUATE CYL 1. FLOW TO CYL 2 IS BLOCKED BY THE SPOOL.

WHEN CYL 1 REACHES ITS LIMIT, THERE IS NO PLACE FOR THE FLUID TO GO, SO PRESSURE INCREASES IN THE PILOT LINE. THIS MOVES THE SPOOL AND SWITCHES FLUID FLOW TO CYL 2.

this function, a **counterbalance valve** such as that shown in Figure 7-16 may be employed. In an actual system, a **counterbalance valve** is installed following a **directional control valve**, which governs the **raising** and **lowering** of the load. When the directional control valve is shifted to **raise** the load, fluid flows into the counterbalance valve, past a **check valve** inside the housing, and out to the **cylinder**.

Upon reaching the full "up" position, **pressure** builds up in the line, and unused fluid is drained back to the **tank**. The **compression spring** in the **counterbalance valve**, however, is adjusted so that the **system pressure** alone is **not** enough to move the spool. The check valve then **closes**. Fluid is therefore **trapped** in the **counterbalance valve**, and the load is **supported**. When the **directional control valve** is shifted to **lower** the load, it directs flow to the **upper port** (the cap end port in Figure 7-16). In pushing down on the piston, it generates pressure in the system which is **added** to the pressure caused by the **load**. This **additional** pressure is enough to lift the counterbalance valve spool. Fluid can now flow back through the spool to the directional control valve.

The spring compression is normally adjusted for a pressure about **10% above** what the load alone produces. It must, of course,

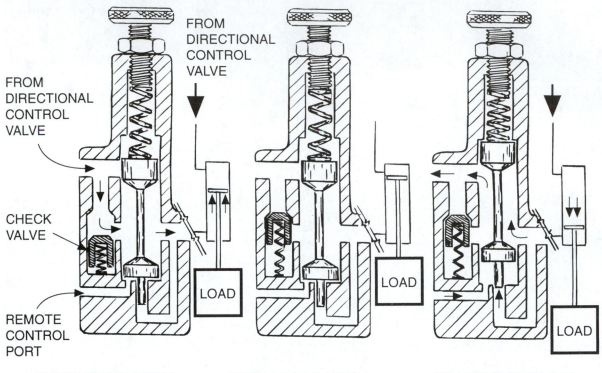

FROM
DIRECTIONAL
CONTROL
VALVE

FROM
DIRECTIONAL
CONTROL
VALVE

CHECK
VALVE

REMOTE
CONTROL
PORT

LOAD

LOAD

LOAD

CHECK VALVE OPENS
TO ALLOW FLUID FLOW
TO RAISE THE LOAD.

WITH LOAD RAISED,
CHECK VALVE AND
SPOOL BLOCK RETURN
FLOW. SPRING IS SET
FOR PRESSURE
SLIGHTLY HIGHER
THAN THE LOAD ALONE
CAN DEVELOP.

TO LOWER THE LOAD,
FLUID IS DIRECTED TO
THE TOP OF THE
CYLINDER OR THE
BOTTOM OF THE
SPOOL, TO OVERCOME
THE SPRING.

Figure 7-16 Counterbalance
valve

be **reset** whenever the weight of the load is **changed**. Pilot pressure
at the remote control port is exerted on a **larger surface** than pressure from the **cylinder**. This means that the spool can be shifted
remotely with **less** pressure.

There are other kinds of pressure control valves used in hydraulic systems, but their operating principles are similar to those
described here. In fact, some manufacturers produce "all-purpose"
valve housings which can be assembled in different ways to serve
different functions. Now that you understand how pressure is used
to cause a valve to shift, and how piloting is accomplished, you
should find it easy to apply this knowledge in learning about other
types of valves.

The purpose of a directional control valve is to either per-mit or block flow within a passage (line), or route flow to or from selected passages in a system.

The **simplest** type of directional control valve allows flow in only **one** direction. This is called a **check valve.** You have already seen check valves used within other types of valves to prevent reverse fluid flow. In every case, a **moving element** within the valve is **seated** to **block** flow in one direction, and **lifts off** the seat to permit flow the other way. In most, a **ball,** usually held seated by a **light spring** is the moving element. In others, the ball is replaced by a **machined poppet.** This usually makes the valve more compact, and when seated, can block two or more passageways, as is illus-trated in the **unloading valve.**

Check valves are available in several **forms.** Some are made for installation within a **straight** run of piping, others for a **right-angle** change in direction. The **swing-gate** valve offers the least **resis-tance** to flow, but must be installed so gravity or a light spring seats the gate to prevent reverse flow. The **spring** may not be needed if the hinge is at the **top** so the **weight** of the gate itself provides proper seating. Sometimes a check valve is fitted with a somewhat **heavier spring** than usual, and inserted in the **return line** to the **tank.** This maintains a **minimum pressure level** which can be used for **piloting** where needed. In such cases the check valve becomes a **pressure-regulating** device. Some applications call for a check valve which can be opened to **reverse** flow. A **pilot-operated check valve** provides this capability. Relatively **low** pres-sure in the **pilot** line, just enough to compress the **light spring** in the valve, controls flow at **high pressure** in a working line. This is a valuable aid in troubleshooting, and could even substitute for a **counterbalance** valve in some cases. Figure 7-18 illustrates its operation.

Similar to the check valve is the **shuttle valve.** This consists of a housing with **two input** ports and **one output** port, and an inter-nal poppet, or **shuttle,** which allows flow from only **one** input at a time.

When a light spring is used to hold the shuttle to one side, as shown in Figure 7-19, the valve is said to be **biased,** and the input ports are identified as **primary,** and **secondary** or **alternate.** With-out the spring, the shuttle remains where **last driven** by fluid flow. It then can be considered a **double check valve.**

While fluid flow in each of these valves is **blocked** or **changed** by an internal moving **part,** the placement, or **position,** of that part is not subject to outside control. All such valves, including **pressure control** valves, are called **one-position** valves. Directional control

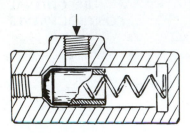

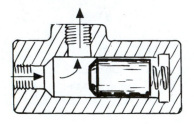

REVERSE FLOW IS
BLOCKED BY A
SPRING-LOADED BALL
OR MACHINED POPPET
(SHOWN HERE).

FORWARD FLOW
COMPRESSES THE
SPRING, OPENING THE
VALVE.

SIMPLE CHECK VALVE

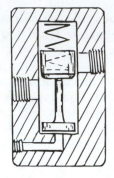

PILOT →

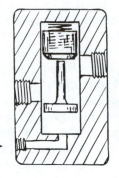

PILOT LINE PROVIDES
THE OPTION OF OPENING
THE VALVE TO REVERSE
FLOW WHEN
NECESSARY.

**PILOT-OPERATED
CHECK VALVE**

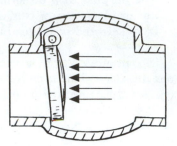

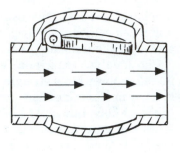

STRAIGHT-THROUGH
FLOW REDUCES
PRESSURE DROP, BUT
VALVE MUST BE
INSTALLED CORRECTLY
OR THE GATE WILL NOT
CLOSE.

**SWING GATE
CHECK VALVE**

Figure 7-17 Check valves

valves are available with up to **six** positions, but **two-position** and
three-position are most common.

"Position," "way," and "port" were defined in Chapter 4. Briefly,
a **position** is a specific valve **setting**, a **way** is a **flow path** through
the valve, and a **port** is a plumbing **connection**. **Positions** establish
flow paths, each **way** needs at least one **port** through which fluid
can **exit** the valve, but sometimes **more** than one **port** is provided
for a single **way**. For example, it is common to have **two ports**
draining to the **tank**, but this path is considered only one **way**.

Symbols used on drawings to represent directional control
valves consist of from **one** to **six** connected **boxes**, each represent-

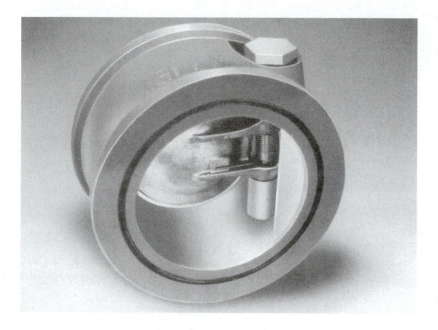

Figure 7-18 Wafer type swing check valve. *Courtesy of Kennedy Valve, Elmira, NY*

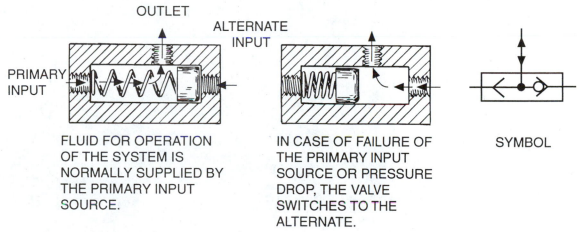

OUTLET

ALTERNATE
INPUT

PRIMARY
INPUT

FLUID FOR OPERATION
OF THE SYSTEM IS
NORMALLY SUPPLIED BY
THE PRIMARY INPUT
SOURCE.

IN CASE OF FAILURE OF
THE PRIMARY INPUT
SOURCE OR PRESSURE
DROP, THE VALVE
SWITCHES TO THE
ALTERNATE.

SYMBOL

Figure 7-19 Shuttle valve

ing a valve **position**. The **group** of boxes is called an **envelope**. Lines extend from one box—the box in the **normal** position if there is one—to represent fluid **conductors** (piping) and show con**nections** to other circuit **components**.

Lines **inside** the boxes indicate **flow paths** and **blocked** passageways for the various **positions**. **Arrowheads** on the flow path lines indicate the **direction** of flow. Lines ending with a crossbar ("T") indicate blocked, or **closed** paths. When the lines inside a position box are all **connected**, in the form of an "**H**," it means all lines are **open** to each other.

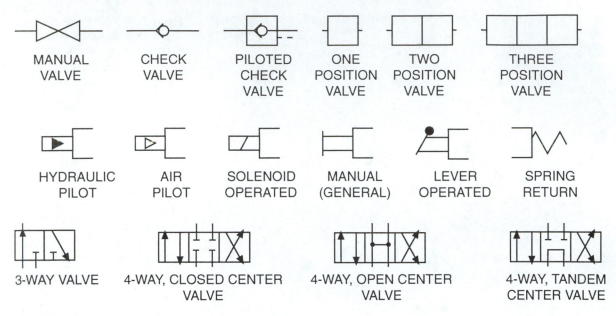

MANUAL VALVE   CHECK VALVE   PILOTED CHECK VALVE   ONE POSITION VALVE   TWO POSITION VALVE   THREE POSITION VALVE

HYDRAULIC PILOT   AIR PILOT   SOLENOID OPERATED   MANUAL (GENERAL)   LEVER OPERATED   SPRING RETURN

3-WAY VALVE   4-WAY, CLOSED CENTER VALVE   4-WAY, OPEN CENTER VALVE   4-WAY, TANDEM CENTER VALVE

Figure 7-20 Examples of valve symbols

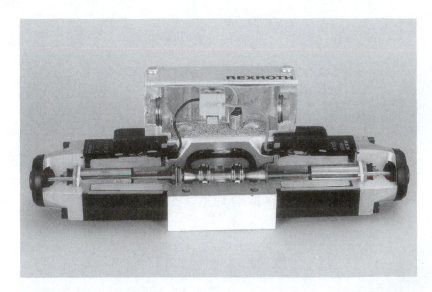

Figure 7-21 Spool valve. *Courtesy of Rexroth Corporation, Bethlehem, PA*

Symbols on the position boxes at each **end** of the envelope indicate how the valve is **operated** (see Figure 7-20). When analyzing circuit operation, you must **visualize** the envelope as a **moving** symbol, but with the **connecting** lines remaining in place. As the envelope moves, **new** connections are made and flow paths **change**. The number of **ways** is usually the same as the number of **lines** leading from the **normal position** box.

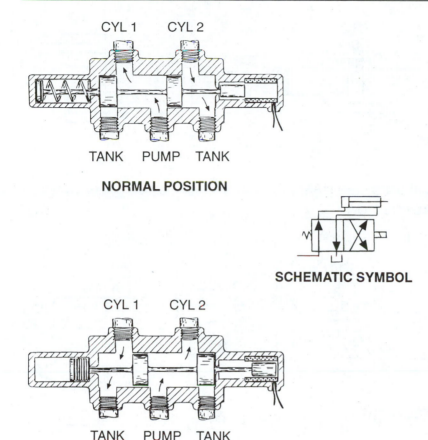

Figure 7-22 Operation of a two-position, four-way, solenoid-operated, spring-return directional control valve

CYL 1   CYL 2

TANK   PUMP   TANK

NORMAL POSITION

SCHEMATIC SYMBOL

CYL 1   CYL 2

TANK   PUMP   TANK

ACTUATED POSITION

Three-way valves are most commonly used where an actuator needs to be driven in one direction only, with the return movement accomplished by gravity or other means, as in a lifting device. One port is connected to the **actuator,** one to the **pump,** and the third to the **tank.** A valve port connected to an **actuator** is called a **working** port. As used in most applications, a **three-way** valve has **one** working port; a **four-way** valve has **two.**

Four-way valves are used when an actuator needs to be driven in **both** directions. The two **working** valve ports are connected to ports at the **cap** and **rod** ends of a cylinder, for example, and are alternately **pressurized** and **exhausted** to the tank. In such applications, the valve may have a **tandem** center position, in which ports are **closed** on one side and **connected,** or **open** on the other. The **closed** ports are connected to the **cylinder** ports, thereby holding the cylinder in place much like a **counterbalance valve.** The

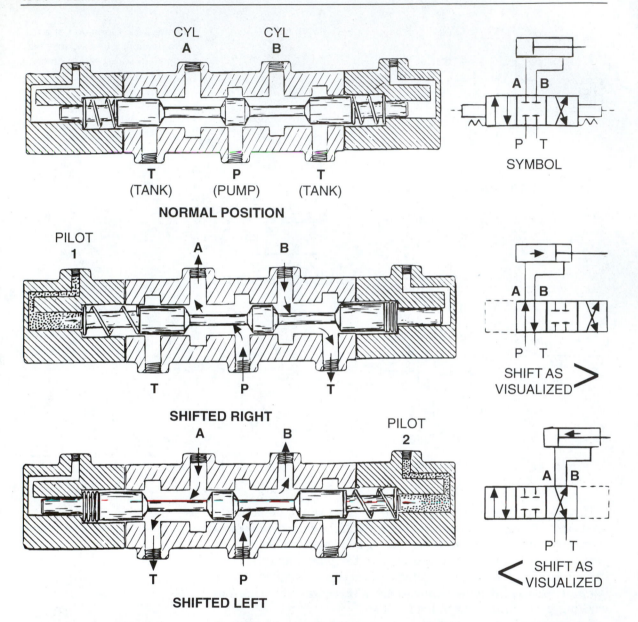

CYL A   CYL B

SYMBOL

T (TANK)   P (PUMP)   T (TANK)

NORMAL POSITION

PILOT 1

A   B

SHIFT AS VISUALIZED

T   P   T

SHIFTED RIGHT

PILOT 2

A   B

SHIFT AS VISUALIZED

T   P   T

SHIFTED LEFT

Figure 7-23 Operation of a three-position, four-way, closed-center, pilot-operated, spring-centered directional control valve

open ports direct flow from the pump directly to the **tank** like an **unloading valve.** This cannot be done, of course, when other actuators are on line with the same pump.

Directional control valves are often assembled together in a single **block,** called a **manifold,** both for convenience in operation and to reduce the likelihood of **leaking connections** or **bursting lines.** You may have seen such manifolds in the cabs of backhoes or bulldozers, or at the operating console of industrial equipment.

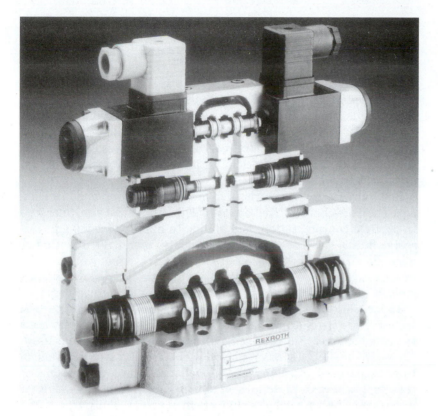

Figure 7-24 Several valve functions combined in a single assembly. *Courtesy of Rexroth Corporation, Bethlehem, PA*

This has been a brief exposure to only a **few** of the seemingly endless variety of valves used in industrial hydraulic installations. You should now be aware, however, of the **principles** and **techniques** involved in **controlling** and **directing** fluid power and be able to apply them to practical problem solving.

CHAPTER SUMMARY

Valves serve **three** specific functions: (1) **control flow**, (2) **control pressure**, and (3) **direct flow**. Whenever the **position** of a valve is **changed**, the valve is said to be **operated**, or **actuated**. If a valve returns to a certain position whenever at rest, or when the system is not **operating**, that is said to be its **normal** position. The means by which valves are **operated** are categorized as **manual, mechanical, pilot**, or **electrical**.

The purpose of a **flow control** valve is to regulate the **amount** of fluid flow, usually to govern the speed of actuators. The simplest flow control valve is the **gate valve**. It is meant to be used **fully open** or **fully closed**. Gate valves are suitable for applications involving high **pressure** and high **volume**, and can tolerate high **temperatures**. However, they are **not** intended for **partially open** operation because they tend to set up **vibration** in the flow. Operation is relatively slow. The fluid used must be **clean,** because any small particle caught between the gate and seat can result in **leakage.**

Globe valves are suitable for high **pressure** applications and **can** be used in the partially open position for **metering** or **throttling.** Flow is intended to take place in **one direction** only, as indicated by an arrow on the **housing.** Fluid must make sharp turns in passing through the **globe** valve, causing pressure drop and turbulence. Fluid must be **clean** in a system using globe valves, as particles can cause leaks.

Ball valves provide control with **low** pressure drop and low **leakage.** They are characterized by low **weight** and small **size,** tolerance to **particle** contamination, and ability to handle **viscous** and **corrosive** fluids. They **open** and **close** quickly, with a 90° rotation of a handle. Their disadvantages are high **cost,** tendency of the seating material to **squeeze** out of place, and the tendency to cause sudden **pressure variations** from quick opening and closing.

Taper plug valves are very **similar** in most respects to **ball valves.** They are **cheaper,** but this advantage may be offset by their tendency to **bind.** For **precise** flow control, a **needle valve** is used. These are available in combination with a **check valve** for applications requiring controlled flow in **one** direction, and **unrestricted** reverse flow. **Pressure-compensated** valves **automatically** maintain a constant flow rate despite changes in **pressure.** The purpose of a **pressure control valve** is to **regulate** fluid **pressure,** either for **safety** or for proper **actuator operation.** Pressure control valves are classified as (1) **simple,** or **direct-acting, and** (2) **compound,** or **piloted.** The purpose of a **pilot valve** is to define fluid flow for control of **valve elements.**

An **unloading valve** is used in systems where the energy lost and heat generated in pressure relief valves is a problem. This is a normally closed type of **compound** valve which, when actuated, develops little back pressure as it sends unused fluid directly to the tank. A **sequence valve** routes oil to **two** cylinders in a specific **order,** for applications which require that the **first** complete its function before flow is directed to the **second.** A **counterbalance valve** is used to **lift** and then **support** a load by means of **trapped** fluid, until a directional control valve is shifted to lower it. The purpose of a **directional control valve** is to either **permit** or **block** flow within a passage, or **route** flow to or from selected passages. A **check valve** is the simplest type of directional control valve. Check valves normally allow flow in only one direction; however, **piloted** check valves have a control passage by which they can be opened to **two-way** flow when necessary. A **shuttle valve** has **two** input ports but only **one** output, and allows only one input at a time. Its purpose is to provide **alternate** sources for an actuator, but only **one** at a **time.**

A **position** is a specific valve **setting.** A **way** is a **flow path.** A **port** is a plumbing **connection.** The number of **ways** is usually the same as the number of **lines** leading from the **symbol** on a drawing. **Three-way** valves are most commonly used where an actuator needs to be driven in **one** direction. **Four-way** valves are used in cases where the actuator needs to be driven in **both** directions. Directional control valves are sometimes assembled in a single **block,** called a **manifold,** for convenience in operation and to reduce **leakage.**

PROBLEMS

7.1 Describe the three specific functions served by valves in a hydraulic system.

7.2 What is meant by the term "normal position" as it applies to a valve?

7.3 Name the four basic ways in which a valve position is changed.

7.4 Explain, with a sketch, how a solenoid operates.

7.5 What is the purpose of a flow control valve?

7.6 List three shortcomings of gate valves.

7.7 Explain one situation in which you would choose a gate valve over a globe valve, and one in which you would choose a globe valve over a gate valve when assembling a system.

7.8 List three advantages and three shortcomings of ball valves.

7.9 What type of valve would you use to achieve precise control over the rate at which a piston rod extends?

7.10 What term is used to describe a valve which is capable of maintaining constant fluid flow even when system pressure varies?

7.11 What is the purpose of a pressure control valve?

7.12 In a pressure relief valve, what term is used to describe the point at which the ball or poppet begins to lift off its seat and cause leakage?

7.13 Explain, with a sketch, what is meant by "chatter" in a pressure relief valve.

7.14 What is the function of a "pilot" valve?

7.15 Since the output of a pressure-compensated pump is automatically adjusted to meet system needs anyway, why do designers sometimes add a compound pressure relief valve?

7.16 Give two reasons why compound, rather than direct-acting, pressure relief valves are not used in all systems, even though they perform better?

7.17 What kind of valve is used in a system to reduce energy waste and heat loss, when a large amount of fluid must be drained to the tank during operation?

7.18 In a system with only one actuator, why couldn't a pressure reducing valve be used in place of a pressure relief valve to limit system pressure?

7.19 Make a sketch, using the proper schematic symbols, showing how a two-position, three-way valve could be used in place of a shuttle valve.

7.20 Indicate the type of each valve named below, using **F** for Flow Control, **P** for Pressure Control, and **D** for Directional Control:

_____Pressure Relief	_____Globe
_____Three-Way	_____Pilot Check
_____Gate	_____Pressure Reducing
_____Check	_____Unloading
_____Counterbalance	_____Sequencing

7.21 What is the primary purpose of a counterbalance valve? How does it work?

7.22 Why are directional control valves sometimes assembled in a single block when their functions are related? What is this "block" called?

7.23 Explain, with a sketch if necessary, how a two-position, four-way valve could be modified and substituted for a three-way valve.

7.24 Explain the term "manual override" and why it is useful.

Cylinders

8

The ultimate purpose of an industrial hydraulic system is to perform a needed function by producing **force** and **motion**. These may be exerted in a straight line, called **linear** motion, or in a circle, called **rotary** motion. The output required by **most** systems is **linear,** and is accomplished by one or more hydraulic **cylinders**. In Chapters 2 and 3 we learned the physical **principles** upon which the applied science of hydraulics is based, and how to determine fluid **pressures**, flow **rates**, and cylinder **sizes**. Chapter 4 showed the **assembly** of a basic hydraulic **system**. In **this** chapter we will learn more about **several** ways in which cylinders are constructed, modified, and used in working circuits to perform more **complex** tasks.

CYLINDER TERMINOLOGY

The simplest type of hydraulic actuation is that which requires **one** piston rod to exert force and motion in **one** direction only. A cylinder having a rod extending from only one end is called **single-ended.** In the cylinders shown in Figure 8-1, pressurized fluid is directed into only one port, so it is called **single-acting** even if there is a port at the other end. The **working** stroke may require the rod to **either** extend or retract. The **return** is most often accomplished by either **gravity** or a **spring.**

If the piston rod extends from **both** ends of the cylinder, it is said to be **double-ended.** It may be either **single-acting** or **double-acting,** depending upon whether the piston is driven hydraulically in **one** or **both** directions. Usually the **force** exerted is **equal** in both

Figure 8-1 Single-acting cylinders

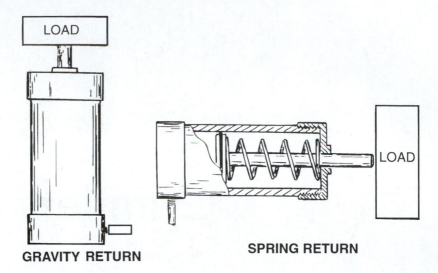

GRAVITY RETURN

SPRING RETURN

directions, since the rod takes up part of the piston area on **both** sides. The forces may be made **unequal,** if desired, by using different sized **rods** or separate **pressure reducing** valves.

TANDEM CYLINDERS

When space available limits the cylinder **diameter** in a particular application, it is often possible to increase the **force** exerted using a **tandem cylinder. Two** pistons on one **rod** are used to increase the effective **area** acted upon by the hydraulic fluid. Tandem cylinders may be either single- or double-acting, and single- or double-ended. Figure 8-2 shows the single-ended type. While the smaller **diameter** of this type makes possible the installation in a **narrower** space, it is more than **twice** as **long** as an ordinary single-piston cylinder.

Figure 8-2 Single-ended tandem cylinder

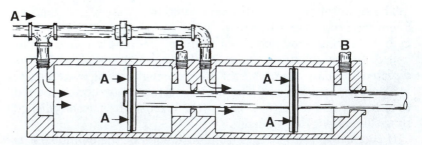

OUTPUT FORCE ON THE ROD IS EQUAL TO THE PRESSURE INPUT AT **A** TIMES THE TOTAL AREA OF BOTH PISTON FACES. FLUID INPUT AT **B** RETRACTS THE ROD, WITH A FORCE EQUAL TO THE PRESSURE AT **B** TIMES THE TOTAL AREA ON THE OTHER SIDE OF BOTH PISTONS.

Wherever conventional cylinders are used, space must be provided for at least **twice** the length of the stroke, since the retracted rod fits **inside** the cylinder. When a cylinder must operate in a **very** limited space, and **length** is the problem, a **telescoping cylinder** such as the one illustrated in Figure 8-3 may be used. Though **expensive** because of their **complex construction, close-fitting parts,** and many **seals,** they satisfy a need which can be met in no other way. In the fully **retracted** position, the piston elements are "nested" inside the main cylinder. They extend, much like a telescope, to convert fluid energy into mechanical force and motion.

TELESCOPING CYLINDERS

Figure 8-3 Telescoping cylinder

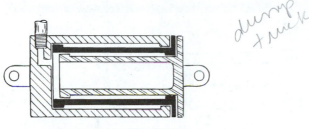

FULLY RETRACTED

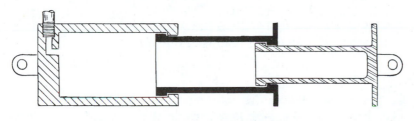

FULLY EXTENDED

It is quite common for a job function to require that some device such as a cutting tool or punch be moved **rapidly** into position with **low** force, then **slowly** at **high** force to do work. These two movements are commonly called the **approach stroke** and the **working stroke**. Within limits, this can be accomplished using a **stepped-piston cylinder.** (Where **very** high force is required, a **pressure intensifier** would be used. These will be covered in Chapter 10).

In a stepped-piston cylinder, a close-fitting **plunger** fits into a **bore** at the cap end when fully retracted. When fluid entering the cap end port **extends** the rod, it acts, at first, only upon the **end** of this **plunger.** Because the bore size is small, the rod extends **rapidly** but with little **force**

STEPPED-PISTON CYLINDERS

$$F = PA.$$

Figure 8-4 Stepped-piston cylinder

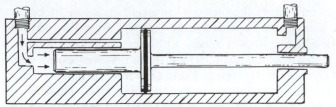

ROD EXTENDS AT HIGH RATE BUT IS CAPABLE OF
RELATIVELY LOW FORCE, AS FLUID ENTERS
SMALLER CHAMBER AND PUSHES AGAINST
SMALLER AREA.

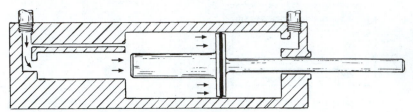

AS FLUID ENTERS LARGE CHAMBER, THE
EXTENSION RATE DROPS BUT THE ROD IS
CAPABLE OF HIGHER FORCE BECAUSE THE
FLUID PUSHES AGAINST A LARGER AREA.

As the plunger **leaves** the bore, incoming fluid enters the **larger** chamber, whose diameter is the inside of the **cylinder**. The force exerted on the rod is now determined by the **pressure** of the fluid and the **total area** of the piston. Notice in Figure 8-4 that the **end** of the plunger **remains** subject to fluid pressure. This simplifies calculations since you need **not** subtract its area in the formula,

$$F = PA.$$

Example: You have an application in which a cutting tool must be moved a total of 4″. During the approach, it must move 3.5″ within 1 second. During the working stroke it must move .5″ with a force of 6000 lbs. The pump is rated at .5 GPM and 400 psi. What diameters of piston and plunger will produce these results?

Solution: The diameter of the piston is determined using two formulas learned in Chapter 2:

$$A = \frac{F}{P}$$

where

$$A = \text{Area of the piston}$$
$$F = \text{Force on the rod}$$
$$P = \text{Pressure of the fluid}$$

$$A = \frac{6000 \text{ lbs.}}{400 \text{ psi}}$$

$$A = 15 \text{ sq. in.}$$

This is the **area** of the piston. We need to know the **diameter**. If given the diameter, we would solve for the area using the formula

$$A = .785d^2$$

where

$$A = \text{Area of the piston}$$
$$d = \text{diameter of the piston.}$$

Here we are given the area, and solve for the diameter using another form.

$$d = \sqrt{\frac{A}{.785}}$$

$$d = \sqrt{\frac{15}{.785}}$$

$$d = \sqrt{19.108}$$

$$d = 4.37'' = \text{Piston diameter (Ans.).}$$

To find the plunger diameter, begin by finding the volume of oil delivered in 1 second at .5 GPM:

$$1 \text{ GPM} = 231 \text{ cu. in./min.}$$

Therefore,

$$.5 \text{ GPM} = 115.5 \text{ cu. in./min.}$$
$$115.5 \div 60 = 1.925 \text{ cu. in./sec.}$$

To move the plunger 3.5" in 1 second, this volume of oil must take the shape of a cylinder 3.5″ long. The formula for finding the **volume** of a cylinder, given the diameter and height, is

$$V = .785d^2h$$

where

V = Volume of the cylinder
d = diameter of the cylinder
h = height of the cylinder.

Here we have the **volume** and height. We solve for the **diameter** using another form.

$$d = \sqrt{\frac{V}{.785h}}$$

$$d = \sqrt{\frac{1.925}{.785(3.5)}}$$

$$d = \sqrt{.7006}$$

$$d = .837″ \text{ Plunger diameter (Ans.).}$$

The working stroke **force** and approach stroke **speed** requirements would be met with a 4.37″ diameter piston and a .837″ diameter, 3.5″ long plunger. Rarely do you find **standard parts** that match precisely the sizes determined through **calculations**. In this case, you would choose a piston size slightly **larger** than the calculated 4.37″ to ensure achieving the 6000 lbs. of **force** needed. A slightly **smaller** plunger would ensure an approach stroke **time** of no more than 1 second.

CUSHIONING

When piston rods operate at **high speed**, especially with heavy **loads,** the sudden impact at the end of a stroke can **damage** either the cylinder or its support. A method used to make the stop more gradual is called **cushioning.** One way of accomplishing this is to **enlarge** a section of the rod on one or both sides of the piston. The enlargement of the **rod** is called a **cushion nose.** An enlarged **end** is called a **cushion spear.** At the end of the cushioned stroke, these fit into a bore called a **cushion cavity,** routing the oil **leaving** the cylinder through a **smaller** passageway. This smaller passage restricts flow, causing a more **gradual** braking at the end of the stroke. Free flow through check valves allows movement at **full** speed during most of the stroke.

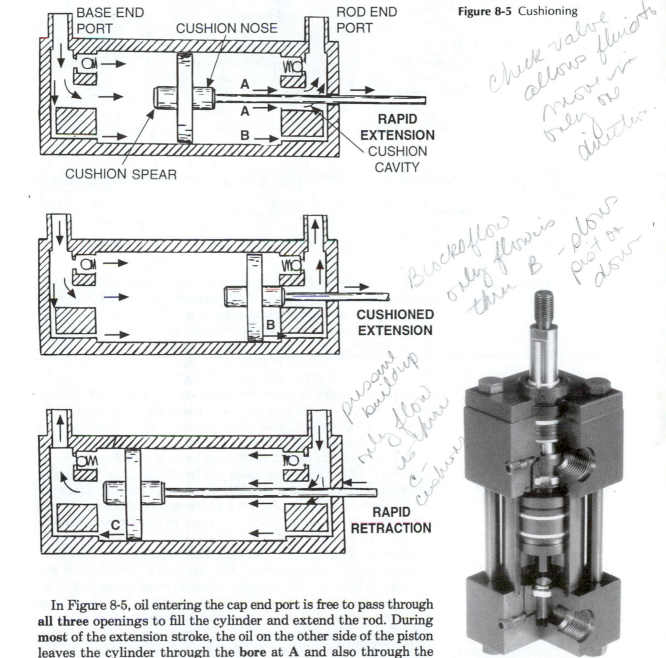

BASE END PORT   **CUSHION NOSE**   **ROD END PORT**

A

A

B

RAPID EXTENSION

CUSHION CAVITY

CUSHION SPEAR

Figure 8-5 Cushioning

Check valve allows fluids to move in only on direction

B

CUSHIONED EXTENSION

Backflow only flows thru B — flows piston to down

pressure buildup only flow is thru C — cushion

C

RAPID RETRACTION

Figure 8-6 Cushioned cylinder with special bearings and seals for high water content fluids. *Courtesy of Vickers, Inc., Troy, MI*

In Figure 8-5, oil entering the cap end port is free to pass through **all three** openings to fill the cylinder and extend the rod. During **most** of the extension stroke, the oil on the other side of the piston leaves the cylinder through the **bore** at **A** and also through the **restriction** at **B**. At the end of the stroke, however, the bore is **blocked** and flow continues only at **B**. The check valve at the rod end is closed. With output flow **limited**, the piston and rod **gradually** brake to a stop.

During **retraction**, flow at the **rod end** passes unrestricted through

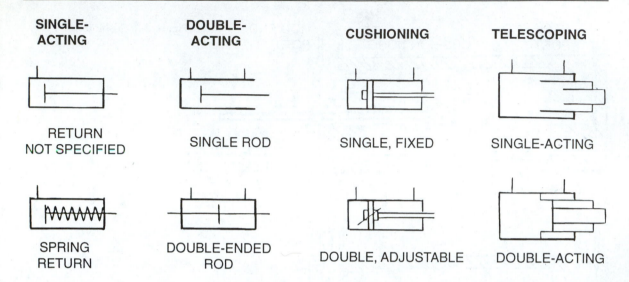

SINGLE-ACTING	DOUBLE-ACTING	CUSHIONING	TELESCOPING
RETURN NOT SPECIFIED	SINGLE ROD	SINGLE, FIXED	SINGLE-ACTING
SPRING RETURN	DOUBLE-ENDED ROD	DOUBLE, ADJUSTABLE	DOUBLE-ACTING

Figure 8-7 Hydraulic cylinder symbols

three passageways, allowing the piston to move at full speed. Cushioning **now** is effective at the **cap end**, as flow is restricted through passageway **C**. Installation of **needle valves** in passageways **B** and **C** would provide an adjustment in the **abruptness** of the braking.

CYLINDER MOUNTING

Several methods for **cylinder mounting** have been devised to suit a wide variety of applications. When a cylinder can be positioned on a flat surface which is oriented **parallel** to the rod motion, either **lugs,** or **tapped holes** in the head and cap may be used. If the rod motion is **perpendicular** to the surface, a **flange** may be attached to either end of the cylinder. The flange is then mounted on the surface with cap screws. Another method involves extending the **tie rods** which hold the cylinder together, then using nuts and washers to assemble the cylinder to a surface. Figure 8-8 shows several typical methods of mounting.

In many applications, the load moves **vertically** or **horizontally** from the piston rod **centerline**. The arms on a backhoe and the blade on a bulldozer are examples. In such cases, the mount must allow the cylinder to move freely to avoid binding of the rod. **Clevis** and **trunnion** mounts provide this flexibility, although on **one axis** only. They may be placed at either **end**, or attached to a block **centered** on the cylinder. You may have seen trunnion mounts on old cannons, on which they provided elevation corrections only.

Rod **binding** is always a potential problem in any installation. If operational requirements dictate a **rigid** assembly, then care must be taken to ensure **precise** positioning so that the rod moves freely, and so repeated cycling or vibration is not likely to disturb

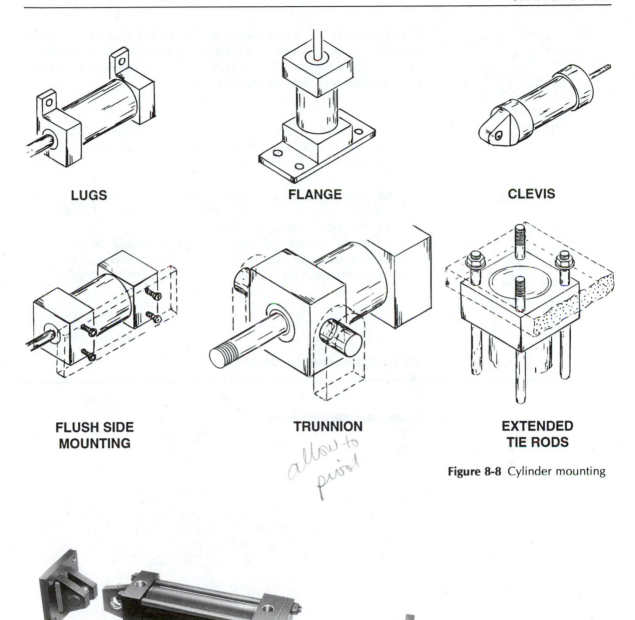

LUGS

FLANGE

CLEVIS

**FLUSH SIDE
MOUNTING**

TRUNNION

*allow to
pivot*

**EXTENDED
TIE RODS**

Figure 8-8 Cylinder mounting

Figure 8-9 Cylinder with
mounting hardware. *Courtesy
of Parker-Hannifin Corp., Fluidpower Group*

the **alignment**. When possible, a **flexible** or self-aligning **coupling** should be used to attach the load to the rod end, to **compensate** for any misalignment. A variety of such devices are provided by manufacturers of fluid power equipment.

STOP COLLARS

Unless the rod is very short, some means of support should be provided to maintain alignment. If a rod is **not** supported, or if **side loading** can occur when extended, there may be some tendency for it to **bind** at the rod end seal. Where **external** support is **not** possible, a **longer** cylinder may be used and a **stop collar** installed. In its simplest form, this is a piece of **tubing** affixed to the rod next to the piston (see Figure 8-10).

The tubing material, of course, must be **compatible** with the fluid used. This means that it will not **dissolve** in or **contaminate** the fluid. Suppliers provide tubing made specifically for this purpose.

Figure 8-10 Stop collar

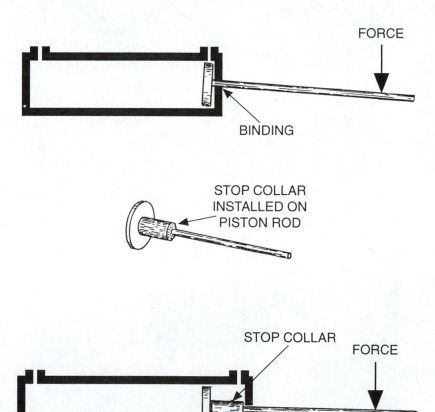

In one sense, perhaps almost **any** group of components assembled to perform a **specific function**, with one or more pressures and flow rates, might be labeled "special purpose." Here, however, we refer to requirements or problems for which somewhat **standard solutions** have been developed. In practice, each is **adapted** to address a specific **need** in system operation. We will review just a few of the **simpler** examples. These are not necessarily the **only** solutions to the problems described, nor perhaps the most appropriate for a specific system.

SPECIAL-PURPOSE CIRCUITS

You have seen how a stepped-piston cylinder produces a stroke with two different characteristics. Initially the rod moves **rapidly**, but is capable of only low **force**. At the conclusion of the stroke, the rod moves **slowly**, but with high **force** potential.

The construction of the stepped-piston cylinder, however, limits its operation to **small** and **medium** flow applications. For **large** cylinders, or tasks requiring two or more cylinders working **together**, a **high-low circuit** as shown schematically in Figure 8-11 may be needed to provide the higher flow rate. **Two pumps** are used, one capable of **high** volume but at **low** pressure, the other having a relatively **low** GPM output but rated for **high** pressure. In practice, this has been found more **efficient** and **economical** than a single, **large** pump having both of these characteristics. In operation, **both** pumps deliver fluid to the cylinder (or cylinders)

HIGH-LOW CIRCUITS

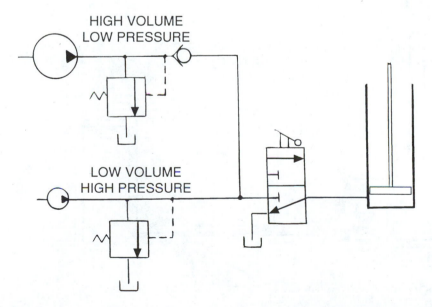

Figure 8-11 High-low circuit

HIGH VOLUME
LOW PRESSURE

LOW VOLUME
HIGH PRESSURE

until the **pressure** developed by a heavy load rises **above** the capability of the **high volume** pump. The motion of the rod is therefore **initially** at **high** speed.

As **high** pressure develops, oil from the high **volume** pump goes back to the tank while the other pump, capable of moving oil at high **pressure,** continues to drive the piston. A **check valve** protects the high **volume** pump from the increase in pressure. When the directional control valve is operated to permit **return** motion, oil from **both** pumps is returned to the tank through pressure relief valves. The limitation on this method is the **pressure** rating of the **low** volume pump. For applications requiring even **higher** pressure, **a pressure intensifier** would be used.

REGENERATIVE CIRCUITS

Sometimes an application requires that a piston rod extend at a **higher** rate than the volumetric output (GPM) of the pump can **generate.** For example, a cylinder designed to exert a force of 1000 lbs. may, at certain times in its operating cycle, need to extend **faster** but with less **force.** One solution is the **regenerative circuit** shown schematically in Figure 8-12. As we learned in Chapter 3, the force **extending** a rod is greater than the force **retracting** it because of the difference in area on opposite sides of the piston. In a regenerative circuit, oil under pressure is directed into ports at **both** ends of a cylinder. The rod **extends,** but is capable of exerting a force equal only to the **difference** between the **extension** and

Figure 8-12 Regenerative circuit

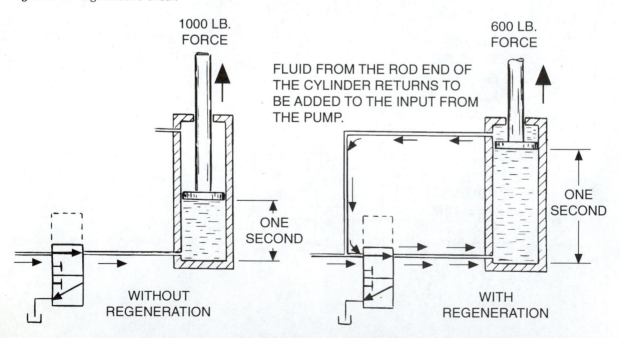

1000 LB. FORCE

600 LB. FORCE

FLUID FROM THE ROD END OF THE CYLINDER RETURNS TO BE ADDED TO THE INPUT FROM THE PUMP.

ONE SECOND

ONE SECOND

WITHOUT REGENERATION

WITH REGENERATION

retraction forces. During extension, oil from the **rod end** port is added to the oil from the **pump**, increasing the total flow **into** the **cap** end. The rod therefore extends at a **faster** rate, but with **less force**.

In an **ideal** system, the pump **output** matches the cylinder **volume** so the rod extends at both the **force** and **speed** needed with **no** oil diverted back to the **tank**. In **actual** circuits, some form of **speed control** must be used when the rate of motion is important. Three types of speed control are commonly used in hydraulic systems:

1. **bleed-off** control, in which part of the oil **bypasses** the cylinder and goes directly to the tank;
2. **meter-in** control, which uses a flow control valve to restrict the flow rate from the pump **into** the **cylinder;** and
3. **meter-out** control, which uses a flow control valve to restrict the flow rate **from** the **cylinder** to the tank.

In the following explanations and illustrations, it is assumed that speed is controlled during rod **extension**.

Bleed-off control is the most efficient of the three, because system **pressure** is determined by the **load** rather than by the setting of the **pressure relief** valve (see Figure 8-13). The pressure drop across the **flow control** valve is the same as the pressure developed by the **load**. This means that the pump need not **work** as hard, and therefore it uses **less energy**. Both the **range** of control and the **precision** with which speed is maintained are **limited** with this method, as compared with the others. Since some of the oil from the pump **always** flows through the flow control valve, the piston can never move as fast as it would with **full** pump flow. Any **change** in system **pressure** changes the amount of oil diverted through

Figure 8-13 Bleed-off speed control

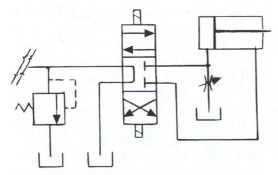

SPEED CONTROLLED DURING
ROD EXTENSION ONLY.

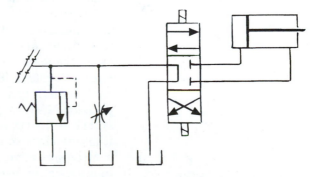

SPEED CONTROLLED DURING BOTH
EXTENSION AND RETRACTION.

Figure 8-14 Speed control by
metering

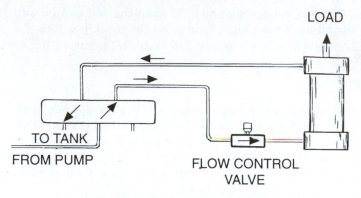

METER-IN SPEED CONTROL

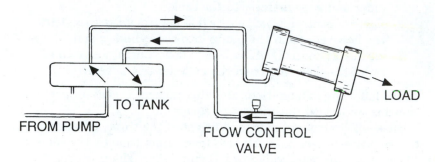

METER-OUT SPEED CONTROL

the **flow control valve**. This changes the amount going into the
cylinder. Pressure variations also alter the amount of **slippage** in
the pump, changing the **output** flow and therefore the **rate** of piston **movement**. Finally, this method does not provide **positive**
control of load movement. If some other force such as gravity
tends to move the load **faster**, it can do so.

Meter-in control is appropriate only when the **load** does not
vary, and **always** offers **resistance** to movement. It may be helpful
to think of this method as controlling the **oil flow** but not necessarily the **load**. Like the bleed-off method, meter-in control is not
positive control. That is, if an external force tends to move the
load **faster**, it can do so. This method is particularly suited for applications such as **lifting** or **compressing** material. System pressure
is that set by the **pressure relief** valve, so efficiency depends upon
its being close to the pressure developed by the **load**.

Meter-out control is used where **both** lower and upper speed
limits must be held in check. The load is **not** free to move faster
when acted upon by external forces. As with meter-in, best efficiency is achieved when the **relief valve** is set to match the pres-

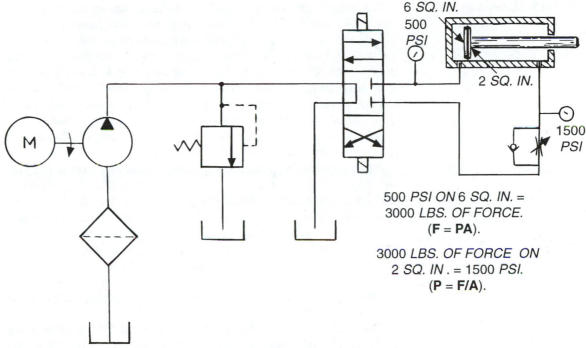

6 SQ. IN.
500 PSI

2 SQ. IN.

1500 PSI

500 *PSI ON* 6 *SQ. IN.* =
3000 *LBS. OF FORCE.*
(**F = PA**).

3000 *LBS. OF FORCE ON*
2 *SQ. IN.* = 1500 *PSI.*
(**P = F/A**).

Figure 8-15 Pressure intensification with meter-out control

sure developed by the **load**. All the lines between the pump and
the cylinder carry fluid at **this pressure**, which means that cylinder
response is **immediate** when the control valve is opened or closed.

When **meter-out** control governs rod **extension**, care must be
taken to avoid creating potentially hazardous **pressure intensification**. This is illustrated in Figure 8-15. The pressure relief valve
located in the pump discharge line limits pressure and provides
protection only up to the **cap end** port of the cylinder. The **force**
on the rod is determined by this inlet pressure and the **total** area
of the piston:

$$F = (P)(A_{piston}).$$

Fluid pressure on the **rod** side of the piston is determined by this
force and the piston area **minus** the **rod** area:

$$P = \frac{F}{A_{piston} - A_{rod}}.$$

This higher pressure exists between the rod end port and the flow
control valve. The **increase** in pressure depends upon the **size of**
the piston **rod**, and the **smaller** the rod, the **lower** the pressure.

CHAPTER SUMMARY

The output of **most** hydraulic systems is **linear,** and is accomplished by one or more hydraulic **cylinders.** A cylinder having a rod extending from only one end is called **single-ended.** If the **working** motion is in one direction only, either **extending** or **retracting** the rod, the cylinder is said to be **single-acting.** If the rod extends from **both** ends of the cylinder, it is called **double-ended.** Usually the **force** exerted by a **double-ended** cylinder is **equal** in both directions, since the rod takes up part of the piston area on **both** sides. The force may be **made** unequal if desired, by using different size **rods** or separate **pressure reducing** valves.

When space available limits cylinder **diameter,** a **tandem cylinder** may be used to achieve the **force** needed. Tandem cylinders may be either single- or double-ended and single- or double-acting. When a cylinder must operate in a very limited space, and **length** is a problem, a **telescoping** cylinder may be used. These are expensive because of their complex **construction, close-fitting** parts, and several **seals.** A **stepped-piston cylinder** provides the capability of variable force and motion. A rapid **approach** stroke with relatively low **force** capability is followed by a slower **working** stroke capable of exerting a **high** force.

When piston rods are required to operate at **high** speeds, especially with heavy **loads,** the sudden impact at the end of the stroke may damage either the cylinder or its support. Therefore, **cushioning** is used to make the stop more gradual. Several methods of **cylinder mounting** have been devised to satisfy a wide variety of applications. When the assembly must be **rigid,** precise **alignment** is critical to avoid rod **binding.** When possible, a **flexible** or self-aligning **coupling** should be used to compensate for any misalignment. Unless a rod is very short, some means of support should be provided if possible. Otherwise, there may be a tendency to bind at the rod end seal. When external support is not possible, one solution is to use a longer cylinder and install a **stop collar** on the rod.

Several special-purpose circuits have been developed and are used to address specific needs in system operation. A **high-low circuit** uses **two** pumps, one capable of high **volume** output and the other capable of meeting high **pressure** requirements. It is used to supply one or more cylinders whose functions require different **approach** and **working** stroke volumes and pressures. When an application requires a rod to extend at a rate **faster** than the pump output can generate, the solution is often a **regenerative** circuit. In this scheme, as the rod **extends,** oil exiting the **rod end** port is directed back to the **cap** end, where it is added to the oil from the pump. The **total** volume of fluid available is increased, thereby extending the rod faster.

In most practical applications where **rate** of rod movement is important, it must be achieved by one of **three** methods of **speed control**. With **bleed-off** control, part of the oil **bypasses** the cylinder and goes directly to the tank. It is the most **efficient** of the three methods, but the **range** and **precision** of control are limited. Furthermore, it does not provide **positive** control of movement, and **external** forces may cause the load to move **faster** than intended. **Meter-in** control is appropriate only when the load is **constant**, and **always** offers **resistance** to movement. It does not provide **positive** control, and **external** forces may cause faster movement than intended. Finally, **meter-out** control holds **both** upper and lower speed limits in check. All the lines between the pump and the cylinder are maintained at **system** pressure, providing **immediate** response when the control valve is opened or closed. When meter-out control is used for rod **extension**, care must be taken to avoid hazardous levels of **pressure intensification**.

PROBLEMS

8.1   What is the term that refers to motion in a straight line?

8.2   What is the term that means a piston rod extends from both ends of a cylinder?

8.3   Name two ways in which the rod of a single-acting cylinder is returned to its normal position.

8.4   Explain, with a sketch if necessary, how force on a rod is increased using a tandem cylinder.

8.5   Why might you use a tandem cylinder in a particular application rather than just select a larger diameter cylinder?

8.6   Under what conditions, or for what reason, would you consider using a telescoping cylinder for a particular application?

8.7   Explain, with a sketch, the operation of a stepped-piston cylinder.

8.8   The rod of a stepped-piston cylinder must advance 3.0″ in 1.5 seconds during an approach stroke, then .5″ with a force of 7500 lbs. The pump is rated at .5 GPM and 600 psi. What diameter of piston and plunger will produce these results?

8.9   The rod of a stepped-piston cylinder must advance 4.0″ in 2.0 seconds during an approach stroke, then .25″ with a force of 12,000 lbs. The pump is rated at .75 GPM and 800 psi. What diameter of piston and plunger will produce these results?

8.10 The rod of a stepped-piston cylinder must advance 5.5″ in 2.5 seconds during an approach stroke, then .2″ with a force of 15,000 lbs. The pump is rated at 1.0 GPM and 750 psi. What diameter of piston and plunger will produce these results?

8.11 The rod of a stepped-piston cylinder must advance 6.0″ in 3.0 seconds during an approach stroke, then .25″ with a force of 9000 lbs. The pump is rated at 1.25 GPM and 750 psi. What diameter of piston and plunger will produce these results?

8.12 In calculating the piston size for a stepped-piston cylinder, you find that a diameter of 3.82″ will produce the force needed. Stock diameters available are 3.75″ and 4.0″. Which would you choose and why?

8.13 In calculating the plunger size for a stepped-piston cylinder, you find that a diameter of 1.22″ will produce an approach stroke fast enough to meet specifications. Stock diameters available are 1.188″ and 1.25″. Which would you choose and why?

8.14 What is the purpose of "cushioning" in a hydraulic cylinder?

8.15 What would be the purpose of having a flow control valve in the restricted-flow passageway of a cushioned cylinder?

8.16 Describe a "stop collar" and explain its purpose.

8.17 Name two mounting devices which permit a cylinder to pivot, or "move up and down" when necessary to avoid binding.

8.18 How can binding be avoided when it is impossible to provide external support to a fully extended rod?

8.19 Under what conditions would it be appropriate to use a tandem cylinder?

8.20 Explain briefly, with a sketch if necessary, how a stepped-piston works.

8.21 Sketch a schematic of a high-low circuit, using proper symbols, and explain its function.

8.22 Sketch a schematic of a regenerative circuit, using proper symbols, and explain its function.

8.23 Explain, with a sketch, the bleed-off method of speed control.

8.24 Explain, with a sketch, the meter-in method of speed control.

8.25 Explain, with a sketch, the meter-out method of speed control.

8.26 Why is the bleed-off method of speed control more efficient than the other two?

8.27 Explain with an example and if necessary with a sketch, how pressure is intensified when the meter-out method of speed control is used.

8.28 Explain briefly what is meant by "positive control" with regard to speed control, and why it is not achieved by the bleed-off or meter-in methods.

8.29 How should the pressure relief valve be set to achieve the best efficiency possible when using meter-in or meter-out speed control?

8.30 Give two disadvantages of using bleed-off speed control as compared with the other two methods.

Accumulators

9

When a working system need **not** run **continuously** to fulfill its function, it is often possible to **store** energy between cycles. This energy is then available to either provide **additional** flow when needed or to **substitute** for the **primary** source in the event of failure. In a hydraulic system, storage of energy in the form of **pressurized fluid** is one function of an **accumulator** (see Figure 9-1). It

Figure 9-1 Hydraulic accumulators. *Courtesy of Parker-Hannifin Corp., Fluidpower Group*

177

is important to note that the extra supply of energy is released in the form of additional **flow**, not higher **pressure**. The **pressure** of the fluid stored in the accumulator is established by the setting of the **pressure relief valve**, as is **system** pressure when no accumulator is installed.

CAUTION: **Whenever servicing any hydraulic system in which an accumulator is installed, be sure to unload the accumulator (that is, release fluid pressure) before taking anything apart!**

APPLICATIONS

Consider a system whose actuator requires **6 horsepower** to perform a task, but operates only **20 seconds** of each minute. The **average** power requirement, since it works only ⅓ of the time, is 2 **horsepower**. With an **accumulator**, the system needs only a 2 **horsepower** energy source operating **constantly** and **storing** fluid during the ⅔ time frame in which the actuator **rests** between cycles.

In a similar application, an accumulator may be used to shorten the **cycle time** of an operation by speeding up the **return** stroke. A pump with low **volume** output but high system **pressure** rating powers the cylinder during the **working** stroke. **Between** cycles, it continues to run and charge the accumulator with oil to be used for the **return** stroke after the **next** working stroke. With little **force** needed for the return, the accumulator "dumps" the oil more **quickly** than it can be moved by the pump, thereby shortening the **overall** cycle time.

Sometimes a system performs a task which uses only a **small** volume of oil, such as clamping or punching. As an **alternative** to starting and stopping the pump repeatedly, an **accumulator** is used. The pump charges the accumulator with enough oil for **several** operations, and when it runs low, a **pressure switch** turns the pump on for recharging.

Sometimes a series of operations in a manufacturing process is such that the **cycle**, once begun, **must** be completed for economical, functional, or safety reasons. Hazardous material or a heavy load, for example, may be transported through a work area. In the event of power failure, energy stored in an **accumulator** is used to continue the cycle long enough to clear the area. In a similar application, airplanes with hydraulically retractable **landing gear** have accumulators for **emergency** use.

Where **shock waves** or **ripples** in the fluid would intefere with proper operation, an accumulator is installed to **smooth out** the flow. This occurs, for example, where **several** cylinders operating at different **times** work close together on **one** working line.

Thermal **expansion** can be a problem, especially in systems operating in high **temperature** environments. Here the function of an accumulator is to absorb any oil **volume** increase and thereby protect lines and components from **overpressure**. The **expansion tank** in home hot water heating systems does this, and is actually a **low-pressure** bladder type accumulator.

Leakage has always been an issue in hydraulic system maintenance. While the need for constant inspection and cleaning is costly and bothersome, a more **serious** aspect is the effect that **loss** of **oil** can have on system **operation**. Sometimes an accumulator is installed solely to maintain fluid level by **replacing** leaked oil.

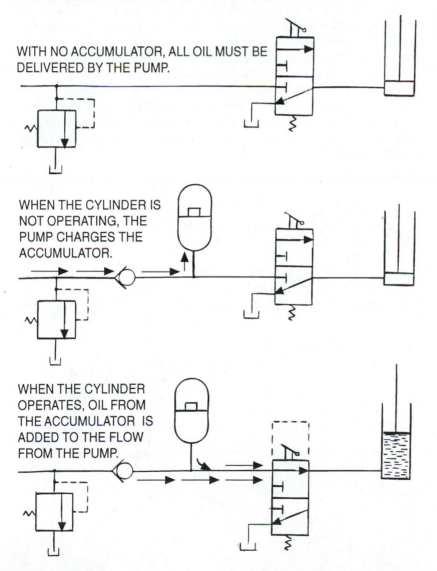

Figure 9-2 Accumulator installation

WITH NO ACCUMULATOR, ALL OIL MUST BE DELIVERED BY THE PUMP.

WHEN THE CYLINDER IS NOT OPERATING, THE PUMP CHARGES THE ACCUMULATOR.

WHEN THE CYLINDER OPERATES, OIL FROM THE ACCUMULATOR IS ADDED TO THE FLOW FROM THE PUMP.

These are just a few examples illustrating the use of accumulators to improve system performance or safety. In most applications, the accumulator is installed in the working line just **ahead** of the cylinder. A check valve prevents **reverse** flow when the accumulator discharges. Figure 9-2 shows one such installation.

ACCUMULATOR OPERATION

With **no** accumulator, **all** oil entering the cylinder must be supplied by the **pump** at its rated GPM **output**. This imposes an easily calculated limit on the **rate** at which the rod extends or retracts. When an accumulator **is** installed in the working line, it becomes a reservoir, or storage place, which the pump fills with oil **between** operating cycles at system **pressure**. Should it become **fully charged** (raised to system pressure by pumped-in oil) **before** the cylinder is again operated, oil from the pump may be directed back to the tank through a **pressure relief** or **unloading** valve (see Figure 9-3).

The next time the cylinder is actuated, the pressurized oil from the accumulator is **added** to the pump flow and moves the piston at a rate determined by the **combined** flow from the **two** sources. The end result is the same as would be achieved using a pump with larger **volume** capability, but the same **pressure** rating.

Figure 9-3 Unloading valve used with an accumulator

WHEN ACCUMULATOR IS FULLY
CHARGED, PRESSURE IN PILOT LINE
OPENS UNLOADING VALVE. OIL
FROM PUMP GOES TO TANK AT
NEARLY ZERO PRESSURE. CHECK
VALVE BLOCKS REVERSE FLOW,
KEEPING ACCUMULATOR CHARGED.

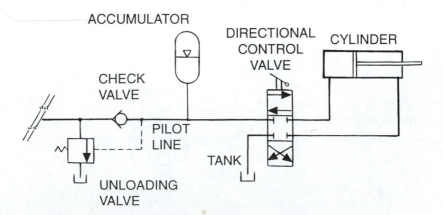

When the function of an accumulator is to provide either an **emergency** source of fluid energy or to **assist** the pump, we first calculate **cylinder** fluid needs using the methods learned in Chapters 2 and 3. If the accumulator is intended for **emergency** use, it must provide **all** the fluid for at least **one** actuation of the cylinder. When **assisting** the pump, it need only be capable of supplying the **difference** between the pump output and the **cylinder** requirement, at or above the **pressure** needed. The specifications for an emergency backup accumulator merely **match** the fluid **volume** and operating **pressure** of the actuator. However, you **may** need to **convert** a cubic inch figure to **liters, quarts,** or **gallons** if your accumulator supplier uses those units.

When the accumulator output will be **added** to that of the pump, the computation requires several steps. First, determine the volume of oil **needed** by the **cylinder** using the diameter and working stroke length. (Remember to **subtract** the **rod** volume if it will be **retracting.**) Next, calculate the volume of oil delivered by the **pump** during the actual **time** of the working stroke from its rated GPM output. The capacity of the accumulator must be at **least** the **difference** between the amount **needed** by the cylinder and the amount **supplied** by the pump during the **time** of the working stroke:

$$V_a = V_c - V_p$$

where

V_a = Volume of the **accumulator,** in cubic inches.
V_c = Volume of oil **needed** by the **cylinder** during actuation.
V_p = Volume of oil **moved** by the pump **during** actuation.

Example: You have a pump rated at 2.0 GPM in a system with a 3″ cylinder. Once every minute the rod of the cylinder must extend 6″ within 2 seconds. How many cubic inches of oil must be supplied by an accumulator? **First** find the volume of oil **needed** by the cylinder to extend the rod 6″.

$$V_c = .785d^2h$$
$$V_c = .785(3^2)(6)$$
$$V_c = .785(9)(6)$$
$$V_c = 42.39 \text{ cu. in.}$$

This is the **total** volume of oil **needed,** and it must be delivered within **2 seconds.** The amount of oil **delivered** by the 2.0 GPM **pump** in 2 seconds, or $\frac{1}{30}$ of a minute is $\frac{1}{30}$ of 2.0 GPM, at 231 cu. in. per gallon. Putting this in the form of an equation you get:

$$V_p = \frac{2(231)}{30}$$

$$V_p = \frac{462}{30}$$

$$V_p = 15.4 \text{ cu. in., in 2 sec.}$$

The volume supplied by the **accumulator** is the **difference** between these two figures:

$$V_a = V_c - V_p$$
$$V_a = 42.39 - 15.4$$
$$V_a = 26.99 \text{ cu. in. (Ans.).}$$

Even if a 27 cu. in. accumulator were available, the one for this application should be somewhat **larger** to allow for the many variations that occur in actual operating systems.

With the exception of the **weighted** type, accumulators constantly **lose** pressure as they return stored oil to the system. Oil is therefore stored at a **higher** pressure than the **load** requires, and as the accumulator empties, it must deliver its **share** before reaching **load** pressure. This means that system pressure, as established by a **pressure relief** or **unloading** valve, will be somewhat higher than it would be without the accumulator.

ACCUMULATOR TYPES

There are three **basic** types of accumulators available, according to the means by which **force** is generated to deliver stored oil to the actuator. These basic types are then subdivided into a variety of constructions to meet specific needs and situations. Those made for industrial applications **generally** range in capacity from **1.5 cubic inches** to **10 gallons**, and have pressure ratings up to **3000 psi**, although higher capabilities are available. The **basic** types are: (1) **weighted**, or **gravity** type, (2) **spring-loaded**, and (3) **gas-charged**. The choice of **type**, **size**, and **installation** for any specific application requires careful consideration of **cost**, **convenience**, **mounting** requirements, **maintenance** considerations, and **functional** characteristics.

WEIGHTED ACCUMULATORS

Although limited in application, **weighted accumulators** offer specific advantages over the other types in those installations for which they are suited. Since the driving force is provided by **gravity**, it is **constant** regardless of the oil level. The oil pressure can

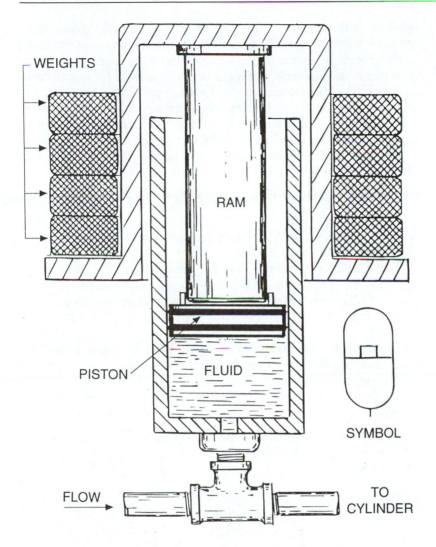

Figure 9-4 Weighted accumulator

WEIGHTS

RAM

PISTON

FLUID

SYMBOL

FLOW

TO CYLINDER

be **changed** simply by adding or removing **weights,** and the weight needed is readily calculated using the formula

$$F = PA.$$

They are typically easy to maintain and service, with little disassembly required. A weighted accumulator is shown in Figure 9-4. Weighted accumulators **must** be mounted **vertically,** and strongly supported. Because of their large **size** and **weight,** and **support** requirements, they are inappropriate for mobile or airborne installations. Their **response** time is slower than that of other types, because of the **inertia** of the weight. For this same reason, they are **not** effective for reducing **pulsations** or **shock waves** in the fluid.

While wide temperature variations can be tolerated, the choice of **material** used in **seals** and **rings** becomes critical for installation in very hot or cold environments. **Leakage** problems, however, are more easily attended to because their construction makes servicing **convenient**. Calculation of the **size** of accumulator to use and the **weight** needed involves only the formulas learned in Chapters 2 and 3. Accumulator **volume** must at least equal the volume of oil to be **delivered**. The **weight** needed is calculated using

$$F = PA,$$

where F is the **weight**, P is the **oil pressure**, and A is the cross-sectional **area** of the accumulator.

Example: The rod of a 3″ cylinder must extend 8″ in 4 seconds with a force of 2000 lbs. The pump is rated at 2.0 GPM and 500 psi. A 2″ diameter weighted accumulator will be installed in the circuit.

(a) How many cubic inches of oil are needed by the cylinder?
(b) Neglecting friction, what pressure is needed?
(c) How much oil must the accumulator supply?
(d) Neglecting friction, how much weight will be needed on the accumulator?

1. To find the amount of **oil needed** by the cylinder, calculate the **volume** to be filled using the **stroke** length for h in the formula:

$$V_{cyl} = .785d^2h$$
$$V_{cyl} = .785(3)^2(8)$$
$$V_{cyl} = .785(9)(8)$$
$$V_{cyl} = 56.52 \text{ cu. in. of oil (Ans.).}$$

2. The oil **pressure** is calculated using the **force** specified and the piston **area**:

$$A_{cyl} = .785d^2$$
$$A_{cyl} = .785(3)^2$$
$$A_{cyl} = .785(9)$$
$$A_{cyl} = 7.065 \text{ sq. in.}$$
$$P = F/A$$
$$P = 2000 \text{ lbs./7.065 sq. in.}$$
$$P = 283.1 \text{ psi, oil pressure (Ans.).}$$

3. The **accumulator** must supply the **difference** between what the cylinder **needs** and what the pump **delivers** in the actuation time of 4 seconds, or $\frac{4}{60}$ of a minute. At 2.0 GPM (462 cu. in. per minute), in 4 seconds, the pump delivers $\frac{4}{60}$ of 462, or 30.8 cu. in.

$$56.52 - 30.8 = 25.72 \text{ cu. in. (Ans.).}$$

4. The **weight** on the accumulator is the **force** which acts upon the accumulator **piston** to generate the oil **pressure**. We need to first find the piston **area** and then use it in the formula

$$F = PA:$$
$$A_{acc} = .785d^2$$
$$A_{acc} = .785(2)^2$$
$$A_{acc} = .785(4)$$
$$A_{acc} = 3.142 \text{ sq. in.}$$
$$F = PA$$
$$F = (283.1)(3.142)$$
$$F = 889.5 \text{ lbs. of weight (Ans.).}$$

SPRING-LOADED ACCUMULATORS

Sometimes an application requires that an accumulator be installed in a hard-to-reach location, or even totally enclosed permanently in plastic or ceramic. Following failure, it is simply replaced rather than serviced. For such installations, among others, the **spring-loaded** accumulator, like the one illustrated in Figure 9-5, is often used. Other advantages are quick **response** time and the ability to function in any **mounting** position. Springs can be made to provide a nearly constant **force** as oil is delivered, but only over a small range of extension; usually force **decreases** as the spring **extends**. During installation, the spring is partially compressed, or **preloaded** to reduce the amount of incoming oil needed to reach the working pressure level. This reduces the total **length** of accumulator required.

As compared with other types, spring-loaded accumulators are **large** and **heavy** for the amount of oil they deliver. They are selected primarily where a small **volume** of oil is needed, and when the **pressure** is in the **lower** ranges and is **not** likely to need **changing**. Metal springs tend to become **fatigued** and **break** after repeated extension and retraction, so this type is **not** used in applications involving rapid **cycling** or those requiring long **service life**. Capacities, operating characteristics, and descriptions of specific products to meet job requirements are detailed in literature available from accumulator manufacturers.

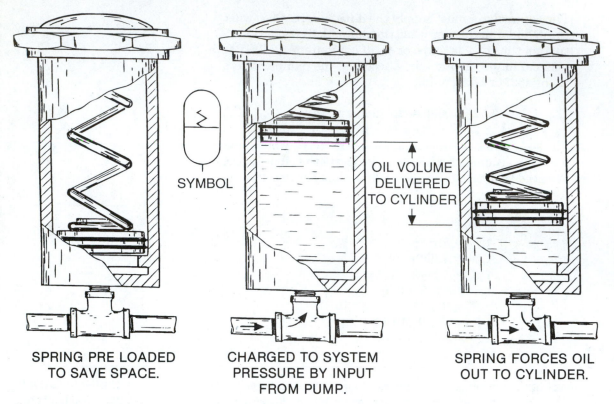

SPRING PRE LOADED
TO SAVE SPACE.

CHARGED TO SYSTEM
PRESSURE BY INPUT
FROM PUMP.

SPRING FORCES OIL
OUT TO CYLINDER.

SYMBOL

OIL VOLUME
DELIVERED
TO CYLINDER

Figure 9-5 Spring-loaded accumulator

GAS-CHARGED ACCUMULATORS

A **gas,** unlike a **liquid,** becomes smaller in **volume** when compressed, or squeezed, and is said to be **resilient,** or "springy." Like a steel spring, it is used as a means of storing **energy** in an accumulator. Just as a **spring-loaded** accumulator is **preloaded** to reduce its size-to-storage ratio, a **gas-charged accumulator** is **precharged** with pressurized **gas.**

Dry nitrogen is the gas preferred for precharging accumulators because it is **inexpensive** and, unlike air, relatively **inert.**

An **inert gas is a fluid with no specific shape** or volume that will **not react with materials to cause oxidation or corrosion.** Helium, **neon, argon, krypton, xenon and radon are inert gases.**

To comply with requirements of the American National Standards Institute (ANSI), dry nitrogen **must** be used wherever pressures exceed **200 psi.**

Accumulators which use pressurized gas to store energy are categorized as **separated** and **nonseparated,** depending on whether the gas **contacts** the fluid. The **nonseparated** type is simply an airtight container with a **gas** valve at the **top** and a **fluid** port at the **bottom**. It must be mounted **vertically** and **motionless** to keep the gas from infiltrating the oil. Such mixing could lead to **cavitation** in the **pump** and **resiliency** (springiness) in the **oil**. Advantages of the nonseparated accumulator are **low cost**, virtually **maintenance-free** service, and the highest possible **storage capacity** for its overall size.

Most gas-charged accumulators are the **separated** type. In these, direct **contact** between the **gas** and **oil** is **blocked** by either a rubberlike material or a piston with seals such as O-rings.

In this type, a rubberlike bag called a **bladder** hangs inside a steel **shell**. A **gas valve** molded into the top of the bladder is used for **precharging** the accumulator with **air** or **dry nitrogen**. Fluid entering the accumulator through the **oil port** at the bottom **compresses** the bladder, storing energy. A **poppet valve** in the port supports the precharged bladder and prevents it from being squeezed into the opening. Figure 9-6 shows the location of these parts.

Although generally limited to **low** and **medium** pressure ranges, the bladder type has special characteristics making it the most **practical** choice for many installations. It is second only to the

GAS BLADDER ACCUMULATORS

Figure 9-6 Bladder type accumulator

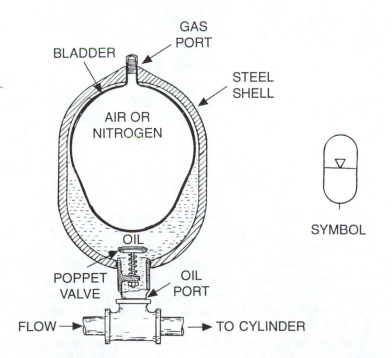

Figure 9-7 Bladder accumulator construction. *Courtesy of Parker-Hannifin Corp., Fluidpower Group*

Figure 9-7 Bladder accumulator construction. *Courtesy of Parker-Hannifin Corp., Fluidpower Group*

nonseparated type in **size-to-capacity** ratio. There can be no **leakage** as long as the bladder remains intact. **Response** to pressure change is **quick**, enabling it to dampen fluid **pulsations**, absorb **shock**, and begin actuator motion **instantly**.

The principal drawback to this type of accumulator is that the bladder material **weakens** after repeated cycling, **bursts**, and needs to be **replaced**. It should not be used where the oil may reach high **temperatures**. Failure is **sudden**, **total**, and **unpredictable**. When failure **occurs**, it cannot be **detected** until the accumulator is called upon to deliver oil, because the pressure gauge will still show system pressure. Usually this means that the **working cylinder** must **malfunction** one or more times before the accumulator failure is **recognized**. For a reasonable balance between best **efficiency** and **long life**, it is generally recommended that bladders be precharged to about **80%** of the **working pressure** required by the cylinder, and that they always be left with about ⅓ of their capacity **after** delivery.

When size and weight are of primary importance, as in an aircraft, the **diaphragm type accumulator** is likely to be the best choice. As Figure 9-8 illustrates, they are assembled in the shape of a **sphere**, or round **ball**, and offer the highest **delivery** capacity for their physical **size** while providing gas and oil **separation**. The diaphragm itself is **preformed** when manufactured, usually in the shape of a pie plate to provide more material and reduce the extent to which it must be stretched. This reduces the strain and makes it last longer.

Diaphragm accumulator characteristics are in most respects **similar** to the **bladder** type, but they are generally available only in capacities up to **one gallon**. Their **response** is **quick**, making them effective for **pulsation** dampening and **shock** absorption and applications where cylinder **movement** must be initiated **promptly**. **Leakage** is zero as long as the diaphragm is **intact**. The rubberlike diaphragm eventually weakens after repeated cycling and bursts. These accumulators are not suitable for applications where oil may reach high **temperatures**. Failure is **sudden, total,** and **unpredictable,** and is usually not detected until the working cylinder malfunctions.

DIAPHRAGM ACCUMULATORS

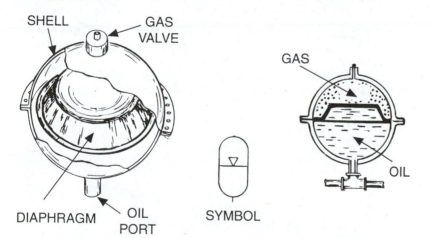

Figure 9-8 Diaphragm type accumulator

The construction of a **piston type gas accumulator** is much like that of a hydraulic cylinder, except that there is no piston **rod** (see Figure 9-9). As with the bladder and diaphragm types, it is precharged through a **gas valve** at the top, and oil enters and leaves through a **port** at the bottom. A free-floating **piston** maintains **separation** between the **gas** and **oil**. For best efficiency, it is generally recommended that the **precharge** pressure in a **piston type** accumulator be about **100 psi** below the **working** pressure required for the cylinder to function. **System** pressure must be **higher** than

PISTON TYPE ACCUMULATORS

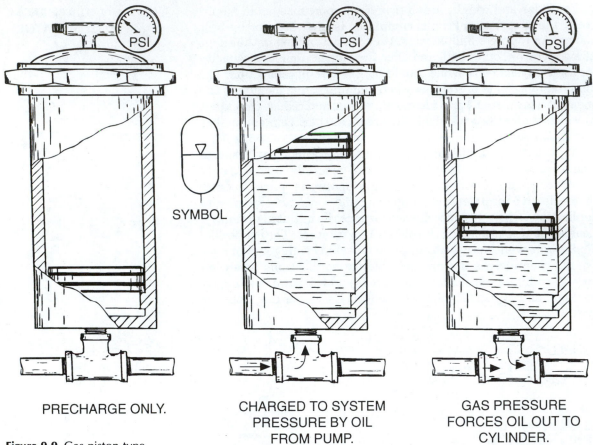

PRECHARGE ONLY.

CHARGED TO SYSTEM
PRESSURE BY OIL
FROM PUMP.

GAS PRESSURE
FORCES OIL OUT TO
CYLINDER.

Figure 9-9 Gas-piston type
accumulator

SYMBOL

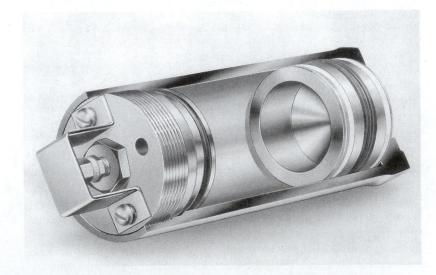

Figure 9-10 Piston accumulator
construction. *Courtesy of Parker-
Hannifin Corp., Fluidpower
Group*

this **working** pressure, since the **force** exerted by the compressed gas **decreases** as oil is delivered.

Piston type accumulators are relatively **expensive**, but they are capable of handling higher **temperatures** and **pressures** than other types. In a system where such conditions exist, the cost is well justified. The **inertia** of the floating **piston** makes this type **unsuitable** for dampening fluid **pulsations** or for absorbing **shock** waves. The possibility of oil **leakage** past the piston seals is always present, especially at high pressures. When used in **low pressure** applications, the loss in **efficiency** due to **force** required to overcome seal **friction** becomes excessive.

As we learned in Chapter 1, for most practical considerations, a **liquid** maintains a constant **volume** despite changes in pressure or the size of its container. A **gas**, however, has **no** specific **volume** and behaves much like a **spring** when enclosed and squeezed. It is this characteristic that makes it suitable in devices such as shock absorbers and accumulators.

When a confined gas is **squeezed**

1. Its **volume** is **reduced**;
2. Its **pressure** is **increased**;
3. Its **temperature rises**.

You may have noticed that a bicycle tire pump heats up when used. Air compressors have cooling fins to increase heat radiation.

If kept at **constant** temperature **during** the compression, or allowed to **cool** to its original temperature **afterward**, the compression is said to be **isothermal**, or "**at one temperature**." Even **without** squeezing, if a confined gas is **heated**, its **pressure** increases as it tries to **expand**. Usually then, when a confined gas is **squeezed**, its pressure increases **first** because of the **compression**, and then even **more** because it is being **heated**.

If it were possible to keep **all** the heat resulting from the squeezing, the compression would be purely **adiabatic**, which means "**without heat gain or loss**." In actual operating systems, compression is **neither** totally isothermal or adiabatic, but something in between. It is easier to learn what takes place, however, if we begin with purely **isothermal** compression.

A **pressure gauge** indicates the **difference** between the pressure of the fluid being **measured** and the **surrounding** air, or **atmosphere**. For gas compression calculations, we must use the difference between the pressure of the measured fluid and **zero**, which represents a **perfect vacuum**. This is called **absolute pressure**, and we refer to it as **psia**. What we **usually** call pressure and refer to as **psi** is, strictly speaking, "gauge pressure," or **psig**.

GAS COMPRESSION

Atmospheric pressure varies with weather changes and **never** gets down to zero. It averages about **14.7 psia** at **sea level** and decreases by about .5 psi with every 1000 ft. of altitude. At **1500 ft.** above sea level, atmospheric pressure stays very close to **14 psia.** In **this** book, we will use 14.0 psia as a reasonably **average** atmospheric pressure, and **convert** psi to psia by adding **14.0** for calculations. For example, **200 psi** converts to **214 psia,** and **1000 psi** converts to **1014 psia.**

If a perfect gas is compressed with **no temperature** change, the **absolute pressure** will **increase** by as many times as the **volume** is **decreased.** If the volume is reduced by **half** its value, the absolute pressure **doubles.** If the volume is reduced to **one-third,** the absolute pressure **triples.** It is expressed in a formula as such:

$$P_1 V_1 = P_2 V_2$$

where

P_1 = Original Absolute Pressure
V_1 = Original Volume
P_2 = Absolute Pressure after compression (working)
V_2 = Volume after compression.

This is known as Boyle's Law and is used for calculating either the **pressure** or the **volume** of a confined gas following **isothermal** compression.

Example: What final gauge pressure results when 200 cu. in. of air at 150 psi is compressed **isothermally** to a volume of 50 cu. in.? First, convert psi to psia:

$$150 \text{ psi} + 14 = 164 \text{ psia}$$
$$P_1 V_1 = P_2 V_2$$
$$164(200) = P_2(50)$$
$$50 P_2 = 32,800$$
$$P_2 = 32,800/50$$
$$P_2 = 656 \text{ psia}$$
$$656 - 14 = 642 \text{ psig (Ans.)}.$$

Example: To what volume must 300 cu. in of gas at 400 psig be compressed **isothermally** to raise the pressure to 1000 psig? First, convert psig to psia:

$$400 \text{ psig} + 14 = 414 \text{ psia}$$
$$1000 \text{ psig} + 14 = 1014 \text{ psia}$$
$$P_1 V_1 = P_2 V_2$$
$$(414)(300) = 1014 V_2$$
$$1014 V_2 = 124,200$$

$$V_2 = 124,200/1014$$
$$V_2 = 122.5 \text{ cu. in. (Ans.).}$$

With purely **adiabatic** compression, the pressure would be **higher** than the figure computed using the conversion formula

$$P_1V_1 = P_2V_2$$

because of the **heat**. The formula needs to be changed for **adiabatic** compression by **increasing** the value of V. This is done with an **exponent**.

You have seen exponents before. In the formula for finding the area of a square,

$$A = S^2,$$

the 2 is an exponent. For purely adiabatic compression of nitrogen, the exponent varies according to pressure and temperature. In general, it is 1.4 for pressures up to **250 psi**, 1.5 for pressures between **250–800 psi**, and 1.6 for pressures between **800–1500 psi**. We will use 1.4 as the "compromise" exponent for our calculations, since no process is **purely** adiabatic and the actual exponent figure **varies**. For **adiabatic** compression, use the equation:

$$P_1(V_1)^{1.4} = P_2(V_2)^{1.4}$$

where

P_1 = Original Absolute Pressure
V_1 = Original Volume
P_2 = Absolute Pressure after compression
V_2 = Volume after compression.

Example: What final gauge pressure results when 200 cu. in. of air at 150 psi is compressed **adiabatically** to a volume of 50 cu. in.? First, convert psi to psia:

$$150 \text{ psi} + 14 = 164 \text{ psia}$$
$$P_1(V_1)^{1.4} = P_2(V_2)^{1.4}$$
$$164(200)^{1.4} = P_2(50)^{1.4}.$$

Using a **scientific calculator** to find the values of $(200)^{1.4}$ and $(50)^{1.4}$:

$$164(1665) = P_2(239)$$

$$P_2 = \frac{164(1665)}{239}$$

$$P_2 = 1142.5 \text{ psia}$$

$$1142.5 - 14 = 1128.5 \text{ psig (Ans.)}.$$

ACCUMULATOR SIZING

An **ideal** accumulator would store **exactly** the amount of oil needed and deliver it to the actuator with the same **pressure** and **flow rate** as a **pump**. With gas accumulators, **both** pressure and flow rate **decrease** steadily as oil is delivered. Furthermore, there are **pressure drops** and **energy losses** due to **friction** as fluid flows through piping and fittings. Whatever the method of calculation, always take these factors into consideration when selecting system **components**, operating **devices**, and pressure **settings**. As a rule-of-thumb, add at least **25%** to the **calculated** accumulator capacity.

To determine the size of **any type** of **gas-charged** accumulator for a specific application, we use two formulas derived from the **adiabatic** compression equation. You will need a scientific calculator.

1. Solve for V_w using

$$V_w = \left(\frac{P_s}{P_w} \right)^{0.714} (V_w - V_{oil}); \text{ and}$$

2. Solve for V_a using

$$V_a = V_w \left(\frac{P_w}{P_a} \right)^{0.714}$$

where

P_a = Precharge pressure
V_a = Volume of the accumulator (gas volume at precharge pressure)
P_s = System Pressure, psi
V_s = Gas Volume at system psi
P_w = Working Pressure needed by the cylinder to function
V_w = Gas Volume at working pressure
V_{oil} = Volume of oil delivered to the cylinder.

Example: You have a system in which a piston type accumulator must deliver 200 cu. in. of oil at 900 psi during each operating cycle. System pressure is 1500 psi and precharge pressure is 800 psi. What size accumulator is needed? First convert **gauge** pressures to **absolute**:

$$P_a = 800 + 14 = 814 \text{ psia}$$
$$P_w = 900 + 14 = 914 \text{ psia}$$
$$P_s = 1500 + 14 = 1514 \text{ psia}$$
$$V_{oil} = 200 \text{ cu. in.}$$

Use these figures to calculate V_w:

$$V_w = \left(\frac{P_s}{P_w}\right)^{0.714} (V_w - V_{oil})$$

$$V_w = \left(\frac{1514}{914}\right)^{0.714} (V_w - 200)$$

$$V_w = (1.656)^{0.714} (V_w - 200).$$

Use the calculator y^x key to raise 1.656 to the 0.714 power:

$$V_w = (1.434)(V_w - 200)$$
$$V_w = 1.434 V_w - 286.8.$$

Add 286.8 to both sides of the equation:

$$1.434 V_w = V_w + 286.8.$$

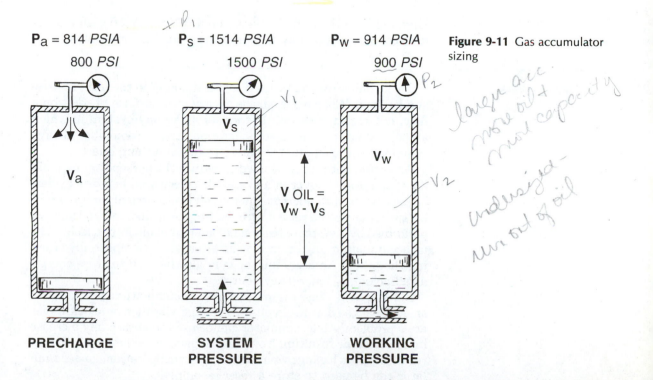

P_a = 814 *PSIA*
800 *PSI*

V_a

PRECHARGE

P_s = 1514 *PSIA*
1500 *PSI*

V_s

V OIL =
V_w - V_s

**SYSTEM
PRESSURE**

P_w = 914 *PSIA*
900 *PSI*

V_w

**WORKING
PRESSURE**

Figure 9-11 Gas accumulator sizing

Subtract 1 V_w from both sides of the equation:

$$.434V_w = 286.8$$

$$V_w = \frac{286.8}{.434}$$

$$V_w = 660.8 \text{ cu. in.}$$

Now calculate V_a:

$$V_a = V_w \left(\frac{P_w}{P_a}\right)^{0.714}$$

$$V_a = 660.8 \left(\frac{914}{814}\right)^{0.714}$$

$$V_a = 660.8 \ (1.123)^{0.714}.$$

Use the calculator y^x key to raise 1.123 to the 0.714 power:

$$V_a = 660.8 \ (1.086)$$
$$V_a = 717.6 \text{ cu. in. (Ans.)}.$$

Dividing by 231, we find this is about 3.1 gal. A **4.0 gallon** accumulator would probably be appropriate for this application.

CHAPTER SUMMARY

An **accumulator** can be used to store **energy** in the form of **pressurized fluid**. When released, the energy is in the form of additional **flow**, not higher **pressure**. Before servicing any system in which an accumulator is installed, it is important to remember always to **unload** the accumulator before taking anything apart.

An accumulator may be used to reduce the **horsepower** capability of the prime mover in a system by **storing** oil between cycles. It may be used to shorten the **overall** cycle **time** of an operation by **speeding up** the **return stroke** of a piston rod. When the task performed by a system uses a **small amount** of oil during each cycle, an accumulator may be used as an alternative to repeatedly starting the pump. An accumulator may also be used to store energy as a **backup**, or **emergency** source.

Bladder and **diaphragm** type gas accumulators are used to **smooth out** the flow in circuits where fluid **vibration** or **shock waves** are a problem. An accumulator can be used to **absorb** any increase in fluid **volume** resulting from heat expansion. Where low oil level resulting from **leakage** would affect system operation, an accumulator can be used to store a **reserve** supply.

When an accumulator provides an **emergency** source of fluid energy, it must be capable of supplying **all** the fluid needed for at least **one cycle** of the actuator. When its output is to be **added** to that of the **pump** during **normal** operation, its size and pressure specifications are determined by the **difference** between cylinder **needs** and pump **output**. With the exception of the **weighted** type, accumulators constantly **lose** pressure as oil is delivered. The oil volume **needed** must be delivered by the time the pressure drops to working pressure. **System** pressure as established by a **pressure relief** or **unloading** valve will therefore be **higher** than it would be **without** the accumulator.

The three **basic** types of accumulators are: (1) **weighted**, or gravity type, (2) **spring-loaded**, and (3) **gas-charged**. **Weighted** accumulators provide **constant** pressure regardless of the oil level. The pressure can be easily **changed** just by adding or removing weight. They tolerate wide **temperature** variations if appropriate **materials** are chosen for seals and rings, and they are typically easy to maintain and service. They must be mounted **vertically**, and are relatively **large** and **heavy**, making them generally unsuitable for **mobile** or **airborne** installations. **Response** is slower than that of other types, making them ineffective for reducing fluid **pulsations** or **shock waves**.

Spring-loaded accumulators are selected primarily for applications where the oil **volume** required is **small**, and the **pressure** is low and not likely to need **changing**. They are especially useful in applications where they must be permanently **enclosed** in plastic or ceramic material, or placed in a hard-to-reach location. They are relatively **large** and **heavy** for their capacity. Since metal springs tend to **fatigue** and **break** after repeated use, they are **not** used where rapid **cycling** or long **service life** are expected.

Gas-charged accumulators are categorized as **separated** and **nonseparated**, depending on whether the gas and oil contact each other. Most used in industry are the **separated** type, using either a rubberlike bag or diaphragm or a piston as a barrier. They are pressurized, or **precharged** with **air** or **dry nitrogen**, usually to about **100 psi** below the **working** pressure, for efficiency.

Gas bladder accumulators, although generally limited to **low** and **medium** pressures, are usually the most **practical** choice for most installations. They have a good **size-to-capacity** ratio, will not **leak** as long as the bladder is intact, and respond **quickly** to pressure change. They are not suitable for high **temperature** applications. When **failure** occurs, it is **sudden, total,** and **unpredictable**. It is generally recommended that bladders be charged only to about ⅔ capacity, and always left with about ⅓ after delivery.

Diaphragm accumulators offer the highest **delivery** capacity for their physical size and weight of any **separated** type. They are capable of quick **response** to pressure change, and are effective in dampening **pulsations** and **shock waves** in the fluid. Like bladder accumulators, they are leakproof as long as the diaphragm remains intact, but failure is **sudden, total,** and **unpredictable.**

Piston type accumulators are constructed much like hydraulic **cylinders,** but without piston rods. They are relatively **expensive,** but handle higher **temperatures** and **pressures** than other types. Their **response** is slow, making them unsuitable for **pulsation** and **shock wave** dampening. The possibility of **leakage** is always present, especially at high pressures. They are relatively inefficient at **low** pressures because of seal **friction.**

When a confined **gas** is **squeezed,** its volume is **reduced,** its pressure **increases,** and its temperature **rises.** If the **heat energy** developed during compression is **removed,** maintaining the gas at a **constant temperature,** the process is said to be **isothermal.** If the **heat energy** is **held** in the gas, causing the temperature to **rise,** the process is said to be **adiabatic.** The formula for **isothermal** compression is:

$$P_1V_1 = P_2V_2$$

The formula for **adiabatic** compression is:

$$P_1(V_1)^{1.4} = P_2(V_2)^{1.4}$$

When selecting system components, it is generally recommended that at least **25%** be added to **calculated** figures to allow for factors such as pressure variations and friction. The calculation of gas-charged accumulator size is based upon the formula for **adiabatic** compression, and requires two formulas:

$$V_w = \left(\frac{P_s}{P_w}\right)^{0.714} (V_w - V_{oil})$$

$$V_a = V_w \left(\frac{P_w}{P_a}\right)^{0.714}$$

PROBLEMS

9.1 Describe briefly four functions of accumulators.

9.2 Name the three basic types of accumulators.

9.3 Describe two advantages and two disadvantages of weighted accumulators as compared with other types.

9.4 Why is a check valve installed in the working line between the pump and accumulator?

9.5 What is the advantage of having an unloading valve rather than a pressure relief valve in an accumulator circuit?

9.6 In what form is energy stored in an accumulator? *pressurized fluid*

9.7 The rod of a 2.5″ cylinder must extend 6″ in 3 seconds with a force of 1800 lbs. The pump is rated at 1.75 GPM and 600 psi. A 1.0″ diameter weighted accumulator will be installed in the circuit.
   (a) How many cubic inches of oil are needed by the cylinder?
   (b) Neglecting friction, what hydraulic pressure is needed?
   (c) How many cubic inches of oil must be supplied by the accumulator?
   (d) Neglecting friction, how much weight will be needed on the accumulator?

9.8 The rod of a 3.5″ cylinder must extend 9″ in 6 seconds with a force of 2200 lbs. The pump is rated at 1.25 GPM and 300 psi. A 1.25″ diameter weighted accumulator will be installed in the circuit.
   (a) How many cubic inches of oil are needed by the cylinder?
   (b) Neglecting friction, what hydraulic pressure is needed?
   (c) How many cubic inches of oil must be supplied by the accumulator?
   (d) Neglecting friction, how much weight will be needed on the accumulator?

9.9 The rod of a 4.5″ cylinder must extend 7.5″ in 5 seconds with a force of 3000 lbs. The pump is rated at 3.0 GPM and 500 psi. A 2.25″ diameter weighted accumulator will be installed in the circuit.
   (a) How many cubic inches of oil are needed by the cylinder?
   (b) Neglecting friction, what hydraulic pressure is needed?
   (c) How many cubic inches of oil must be supplied by the accumulator?
   (d) Neglecting friction, how much weight will be needed on the accumulator?

9.10 The rod of a 6″ cylinder must extend 3″ in 2.5 seconds with a force of 8000 lbs. The pump is rated at 5.0 GPM and 500 psi. A 1.5″ diameter weighted accumulator will be installed in the circuit.
   (a) How many cubic inches of oil are needed by the cylinder?
   (b) Neglecting friction, what hydraulic pressure is needed?
   (c) How many cubic inches of oil must be supplied by the accumulator?
   (d) Neglecting friction, how much weight will be needed on the accumulator?

9.11 What precaution must be taken before servicing any hydraulic system in which an accumulator is installed?

9.12 You have a hydraulic system driven by a 10 horsepower electric motor delivering 3 GPM. Once every minute it extends the rod of a cylinder for 12 seconds, and then rests for 48 seconds. Ideally, what size motor could be used if an appropriate accumulator were installed?

9.13 You have a hydraulic system driven by a 12 horsepower electric motor delivering 4 GPM. Once every minute it extends the rod of a cylinder for 15 seconds, and then rests for 45 seconds. Ideally, what size motor could be used if an appropriate accumulator were installed?

9.14 How would system operation be affected if the precharge pressure on a gas-piston accumulator were higher than the pressure relief valve setting?

9.15 Explain, with a sketch, how an unloading valve improves the efficiency of a system in which an accumulator is installed.

9.16 Why is a weighted accumulator not effective in absorbing fluid vibrations or shock waves?

9.17 Explain briefly how and why an accumulator chosen as an emergency backup for the pump in a system might be different from one installed to permit using a smaller pump.

9.18 If your calculations showed that a 3.95 gallon accumulator would deliver the amount of oil needed by a cylinder, why would it not be appropriate to order a 4.0 gallon capacity accumulator for this application?

9.19 Why might the pressure setting of an unloading valve need to be higher in a system with an accumulator than in a system without one?

9.20 Explain briefly the meaning of the words, "isothermal" and "adiabatic" as they apply to the compression of a confined gas.

9.21 Explain briefly the term "inert gas," and give two examples.

9.22 According to ANSI standards, over what pressure should an inert gas be used in gas accumulators?

9.23 Explain briefly the difference between separated and nonseparated gas accumulators, which type is preferred for most applications, and why.

9.24 What is the recommended guideline to be used in determining what to use as a precharge pressure in a gas accumulator?

9.25 Why is a piston type gas accumulator not usually used in low pressure applications?

9.26 What three things happen to a confined gas when it is squeezed?

9.27 Explain briefly the difference between gauge pressure and absolute pressure.

9.28 What final gauge pressure results when 300 cu. in. of air at 40 psi is compressed isothermally to a volume of 60 cu. in.?

9.29 What final gauge pressure results when 450 cu. in. of air at 70 psi is compressed isothermally to a volume of 45 cu. in.?

9.30 To what volume must 800 cu. in. of air at 60 psig be compressed isothermally to raise the pressure to 90 psig?

9.31 To what volume must 750 cu. in. of air at 80 psig be compressed isothermally to raise the pressure to 200 psig?

9.32 To what volume must 500 cu. in. of air at 100 psig be compressed isothermally to raise the pressure to 200 psig?

9.33 You have a system in which a gas accumulator must deliver 250 cu. in. of oil at 1200 psi during each operating cycle. System pressure is 1800 psi, and precharge pressure is 1100 psi. What size accumulator is needed?

9.34 You have a system in which a gas accumulator must deliver 180 cu. in. of oil at 850 psi during each operating cycle. System pressure is 1200 psi, and precharge pressure is 750 psi. What size accumulator is needed?

9.35 You have a system in which a gas accumulator must deliver 320 cu. in. of oil at 975 psi during each operating cycle. System pressure is 1350 psi, and precharge pressure is 875 psi. What size accumulator is needed?

9.36 You have a system in which a gas accumulator must deliver 270 cu. in. of oil at 1360 psi during each operating cycle. System pressure is 2000 psi, and precharge pressure is 1200 psi. What size accumulator is needed?

9.37 You have a system in which a gas accumulator must deliver 160 cu. in. of oil at 1800 psi during each operating cycle. System pressure is 2500 psi, and precharge pressure is 1600 psi. What size accumulator is needed?

9.38 You have a system in which a gas accumulator must deliver 90 cu. in. of oil at 700 psi during each operating cycle. System pressure is 1500 psi, and precharge pressure is 600 psi. What size accumulator is needed?

Pressure Intensifiers

10

As we learned in Chapter 6, there is an **upper limit** to the amount of system pressure a **pump** can withstand. While the requirements of **most** applications can be met by the **sturdier** models, there are times when the force needed dictates pressures of **several thousand** psi. With a **large** enough cylinder, we could develop **any** size force, even with low pressure. There are, however, practical limits on available **space** and the amount of **fluid** used.

In Chapter 9 we learned how to use storage devices called **accumulators** to increase the **flow rate** to an actuator. In this chapter we will study hydraulic devices used to increase fluid **pressure**, called **pressure intensifiers**, or **boosters**. Units are readily available through commercial suppliers which will multiply fluid input pressures up to 50-to-1, and even greater increases are possible. The **input** may be **either** liquid or gas (air), but the high pressure output is **always a liquid**.

Booster operation is based upon the relationship between **area, force**, and **pressure** which we learned about in Chapter 2. Fluid pressure can be **intensified** using two cylinders of different diameters, **rigidly** mounted in line so the rod of the **first** drives the rod of the **second**. In Figure 10-1, a fluid pressure of **80 psi** on a piston area of **50 sq. in.** develops a force of **4000 lbs.**

$$F = PA$$

PRESSURE INTENSIFICATION PRINCIPLE

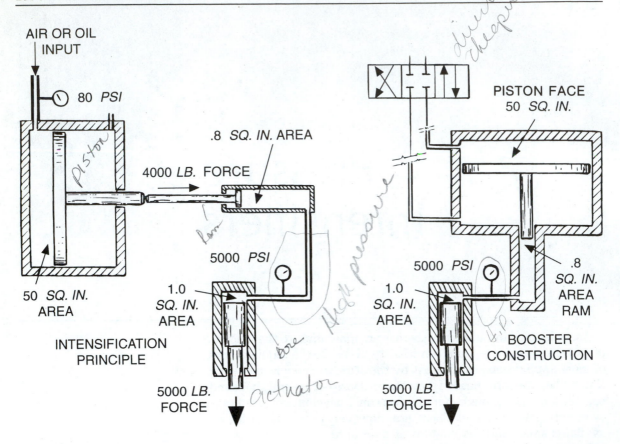

Figure 10-1 Pressure intensification

This **force** is transmitted directly to the rod of a **second** cylinder. The **4000 lb.** force on the **.8 sq. in.** of its piston now develops **5000 psi**.

$$P = F/A$$

The fluid pressure has been increased from 80 psi to 5000 psi, or by **62.5 times**. This would be called an **intensification factor** of **62.5**. In a **booster,** the two rods are combined into one solid piece. Pressure intensification results from the difference in **area** between **piston face** and the **end** of the **ram**.

APPLICATIONS

A booster may be appropriate for any task in which a **small** amount of **high** pressure liquid is needed and a **large** amount of **lower** pressure liquid **or** gas is available. It is also a means by which high pressure can be developed and maintained for long periods with little or no loss of energy. Directional and pressure **control** functions are accomplished at **low** pressure, while **high** pressure fluid is developed only between the **booster** and **actuator**.

Since the increase in pressure is accomplished without the need for any **electrical** power input, boosters present no fire or explosion hazard. The conversion is purely **mechanical** in nature. Also, since the input and output fluids are **separated,** the two can have different **characteristics.** The pressure of a dangerous or corrosive fluid, for example, can be raised using an air or oil input at low pressure. When the required pressure is **higher** than any available **pump** can maintain, a booster may be the **only** solution.

In some circuits, the **highest** pressure level is **within** pump limitations, but is needed in only one **part** of a system. Here, for reasons of **economy, safety,** and **efficiency,** a booster is installed to provide high pressure only where **needed.** The **rest** of the circuitry, at **lower** pressure, can be piped and equipped with less expensive conductors and components. Typically, valves, seals, pressure control devices, and fittings designed for **low pressure operation** are less **expensive** and easier to **install** and **maintain.** With **most** of the conductors and components at **low** pressure, the system itself is **less dangerous** should a line burst or a connection fail. A pressure relief valve passing oil back to the tank at **low** pressure generates **less heat.**

Most factories have compressed air available for a variety of purposes, commonly at about **60–80 psi** and called "**shop air.**" When only a **small** amount of pressurized oil is needed, shop air provides the **input** pressure to a booster. The **booster** replaces the several components of a complete hydraulic system while the **pressurized oil** provides the **hydraulic** capabilities. Furthermore, these units can be made very **compact** and **portable** for situations in which they need to be moved about in a factory, or even taken to remote locations.

Here the pressure level is established by an **air pressure regulator** rather than by a pressure relief valve. Air pressure regulators use no energy, thereby increasing efficiency, and also provide more precise control. Because air is **compressible,** boosters pressurized with shop air provide a somewhat **smoother** application and release of force, which is often an advantage. It is likely that **many** functions performed by complete hydraulic **systems** could be accomplished equally well, and at less expense, by **boosters** driven by **air.**

The **single-acting** booster is the simplest and most common type, and is used extensively in manufacturing and assembly of parts. In **most** applications, including those shown in Figure 10-2, the pressurized oil is directed to an **actuator,** where it is converted into **force** to do work. However, certain **metal-forming** processes use a booster to convert **air** pressure to high pressure **water.** Thin-walled metal cylinders, locked in a **die,** are expanded to the desired shape by the pressure of the water (see Figure 10-3). When

Figure 10-2 Booster applications

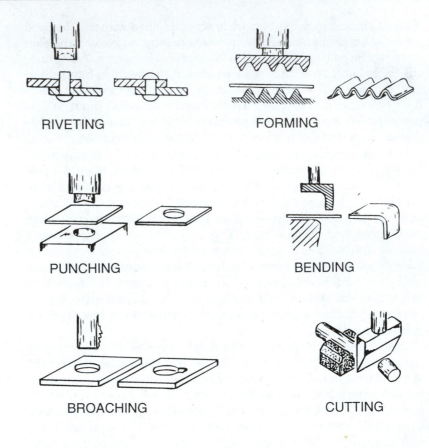

RIVETING   FORMING

PUNCHING   BENDING

BROACHING   CUTTING

Figure 10-3 Metal forming with high-pressure water

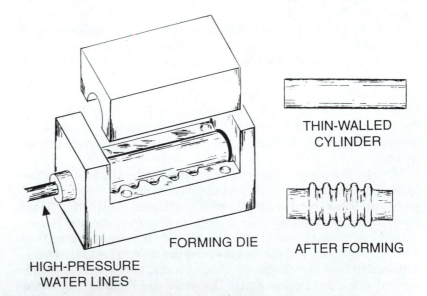

THIN-WALLED CYLINDER

FORMING DIE   AFTER FORMING

HIGH-PRESSURE WATER LINES

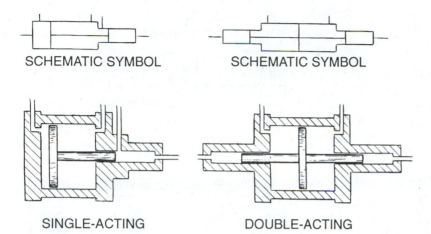

Figure 10-4 Single- and double-acting boosters

SCHEMATIC SYMBOL

SCHEMATIC SYMBOL

SINGLE-ACTING

DOUBLE-ACTING

a **continuous** supply of high pressure fluid is needed, a **double-acting** booster may be used (see Figure 10-4). With each stroke, the flow **alternates** between input ports, **reversing** the direction. **High** pressure fluid flows alternately from the ram end ports, through check valves, to the actuator.

DUAL PRESSURE OPERATION

As we learned in Chapter 3, it takes a **large** volume of fluid at **low** pressure to produce a **small** volume of fluid at **high** pressure. It is **wasteful** to use **high** pressure fluid in any part of an operation where a **lower** pressure would serve the purpose. Some applications require that an actuator move some distance **before** contacting the workpiece. There is little or no resistance to this initial motion, called the **approach stroke**. The low **input** pressure is enough to move the actuator. The **higher**, or **intensified** pressure is needed only for that motion requiring **high** force, which is called the **work** stroke. A **dual pressure** booster as illustrated in Figure 10-5 provides this capability.

Most installations use **shop air** as the source of input pressure, with **advance** and **retract** air-over-oil tanks to hold oil for the booster and actuator. A **directional control valve** pressurizes the two tanks to alternately advance and retract the actuator plunger. When the actuator plunger contacts the workpiece, increasing pressure in the **input** line, a pilot line in a **second** directional control valve is pressurized. This directs air flow to the **piston chamber** of the booster, forcing the ram downward and **cutting off** the flow from the **low pressure** port. Since the surface area of the **piston** is greater than that of the **ram**, the pressure of the trapped oil is now **increased**, or **intensified**.

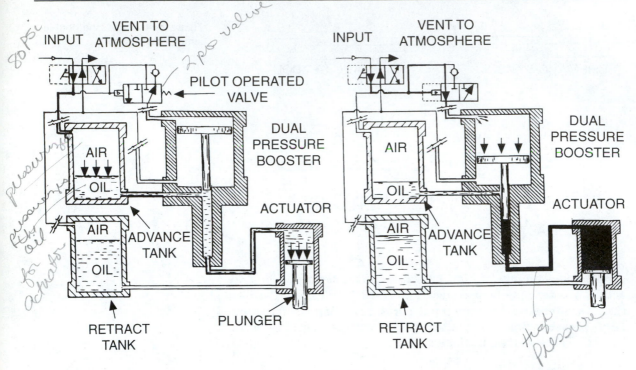

Figure 10-6 Dual pressure ram stroke

OPERATION IS INITIATED BY OPENING A VALVE DIRECTING LOW PRESSURE AIR TO THE **ADVANCE TANK**. AIR PRESSURE ON THE SURFACE OF THE OIL MOVES IT THROUGH THE BOOSTER TO THE ACTUATOR. THE ACTUATOR PLUNGER ADVANCES THROUGH THE **APPROACH** STROKE. AIR PRESSURE IS NOT HIGH ENOUGH TO ACTUATE THE **PILOT-OPERATED** VALVE BECAUSE RESISTANCE OFFERED BY THE PLUNGER IS LOW.

AT THE END OF THE **APPROACH** STROKE, HIGH RESISTANCE **RAISES** THE FLUID PRESSURE AND SHIFTS THE PILOT-OPERATED VALVE. INCOMING AIR NOW ACTS ON THE BOOSTER **PISTON**, ADVANCING THE **RAM**. HIGH PRESSURE OIL (SHOWN IN BLACK) DRIVES THE ACTUATOR PLUNGER WITH **HIGH** FORCE THROUGH ITS **WORK** STROKE.

Although for **most** applications and calculations, liquids are **considered** incompressible, this is not strictly true. Under very **high** pressure, hydraulic oil does compress, or **shrink** slightly. The **exact** amount varies depending on the oil and the amount of entrapped air, but it ranges between **1%–3% per 1000 psi** (see Figure 10-6). This is called the **compressibility factor**. When the oil pressure is **2000 psi or higher,** this must be included when calculating the booster ram stroke **length**.

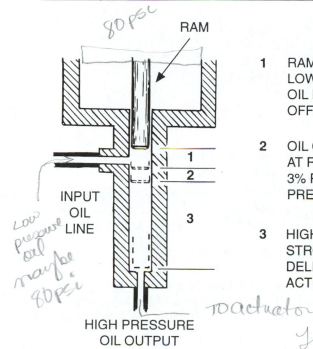

AIR.
80 PSI

RAM

Low pressure oil maybe 80 PSI

INPUT
OIL
LINE

HIGH PRESSURE
OIL OUTPUT

Figure 10-5 Dual pressure booster operation

pre travel of ram

1   RAM PRE-TRAVEL. LOW PRESSURE OIL INPUT IS CUT OFF.

2   OIL COMPRESSES AT RATE OF 1% TO 3% PER 1000 PSI AS PRESSURE BUILDS.

3   HIGH PRESSURE STROKE, AS OIL IS DELIVERED TO THE ACTUATOR.

TO actuator   *low pressure to move actuator. (2 speed similar)*

A **single pressure** booster is used when the actuator must exert high force for the **full** length of its stroke. Selection of a booster for a specific application begins, quite logically, with the calculation of the oil **pressure** needed to drive the actuator and the **volume** needed for it to complete its function. If the intensified oil pressure is **2000 psi** or more, you need to include the **compressibility factor** in your calculations.

Consider a metal forming operation in which a 3″ diameter plunger must exert 9000 lbs. of force through a 5″ stroke. We have **80 psi** shop air available, which will be intensified to generate the force required. Using calculations which assume no **friction** losses, we need to determine:

1. What **intensification factor** (booster ratio) is needed, and
2. What **size** booster would be suitable, considering reasonable piston and ram **diameters** and **stroke** length (see Figure 10-7).

First find the pressure needed on the 3″ actuator to produce 9000 lbs. of force, using

$$P = F/A.$$

SIZING A SINGLE PRESSURE BOOSTER

$$A = .785d^2$$
$$A = .785(3)^2$$
$$A = .785(9)$$
$$A = 7.065 \text{ sq. in.}$$
$$P = F/A$$
$$P = 9000 \div 7.065$$
$$P = 1273.9 \text{ psi needed.}$$

Now find the intensification factor, or **booster ratio** needed to increase the 80 psi shop air to 1273.9 psi:

$$\text{Booster Ratio} = \frac{\text{Output Pressure}}{\text{Input Pressure}} = \frac{1273.9}{80}$$

Booster Ratio = 15.92, or 16.

The pressure increase is proportional to the **areas** of the booster piston and ram, **not** the **diameters**. The **area** of the booster piston must be **16 times** that of the ram. We learned in Chapter 2 that

Figure 10-7 Sizing a single pressure booster

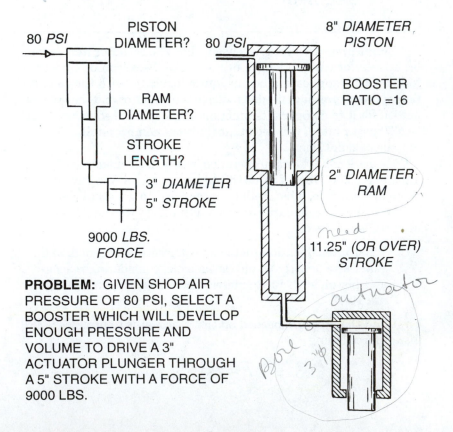

80 *PSI*

PISTON DIAMETER?

RAM DIAMETER?

STROKE LENGTH?

3" *DIAMETER*
5" *STROKE*

9000 *LBS. FORCE*

80 *PSI*

8" *DIAMETER PISTON*

BOOSTER RATIO =16

2" *DIAMETER RAM*

need
11.25" *(OR OVER) STROKE*

PROBLEM: GIVEN SHOP AIR PRESSURE OF 80 PSI, SELECT A BOOSTER WHICH WILL DEVELOP ENOUGH PRESSURE AND VOLUME TO DRIVE A 3" ACTUATOR PLUNGER THROUGH A 5" STROKE WITH A FORCE OF 9000 LBS.

areas are proportional to diameters **squared.** The ratio between the piston and ram diameters is therefore proportional to the **square root** of 16, or 4. The booster piston diameter must be 4 times the ram diameter.

> The ratio between piston and ram diameters is the same as the **square root of the booster ratio.**

Any number of piston and ram sizes would meet this requirement. We could use a 4″ piston and 1″ ram, or an 8″ piston and 2″ ram, or a 12″ piston and 3″ ram, just as examples. Choosing a booster with the **smallest** diameters available saves on compressed air, which is expensive, but requires a long ram stroke to deliver the **volume** of oil needed by the actuator. As with most design decisions, the choice represents a compromise.

Consider a booster having a 4″ piston and 1″ ram. We need to calculate the distance that a 1″ diameter ram must travel to deliver the amount of oil needed by the actuator. Using the formula for the volume of a cylinder to find the amount of oil needed to move a 3″ diameter plunger through a stroke of 5″ you get:

$$V = .785d^2h$$
$$V = .785(3)^2(5)$$
$$V = .785(9)(5)$$
$$V = 35.325 \text{ cu. in.}$$

Now find the **height** of this amount of oil in the 1″ bore of the booster, which is the **length** of the ram stroke. For this we need the formula for the **height** of a cylinder when given the **volume** and **diameter:**

$$h = \frac{V}{.785d^2}$$

$$h = \frac{35.325}{.785(1)^2}$$

$$h = 45″ \text{ stroke length.}$$

This is too long, and it is unlikely that we would find such a booster. We need to consider larger piston and ram diameters. Instead of doing the calculations, however, compare the actuator and ram **stroke lengths** with their **diameters:**

Actuator: 3″ diameter, 5″ stroke
Ram: 1″ diameter, 45″ stroke

The ratio between actuator and ram stroke lengths is the same as the ratio between the squares of their diameters.

In the above example, the ratio between 3^2 and 1^2 is 9-to-1. The ratio between 45 and 5 is 9-to-1.

We now have a simpler way to determine the length of booster stroke for various ram diameters.

For a 1.5″ diameter ram:

$$\text{Stroke} = \frac{3^2}{1.5^2}\,(5) = \frac{9}{2.25}\,(5) = 20''.$$

For a 2″ diameter ram:

$$\text{Stroke} = \frac{3^2}{2^2}\,(5) = \frac{9}{4}\,(5) = 11.25''.$$

A reasonable choice of booster would have an 8″ piston, 2″ ram, and the next available stroke length over 11.25″.

SIZING A DUAL PRESSURE BOOSTER

A **dual pressure booster** is used when the actuator stroke consists of an **approach** stroke which may be powered by **low** pressure followed by a **working** stroke requiring **high** pressure. The booster selection process begins with calculations of the **high** pressure and oil **volume** needed for the **working** stroke alone. If the intensified oil pressure is **2000 psi** or more, you will need to include the **compressibility factor** in your calculations.

Consider a metal cutting operation in which a 3″ diameter plunger must advance a total of 5″, consisting of a 4.5″ **approach** and a .5″ **working** stroke with a **20,000 lb.** force. The high pressure line between the booster and actuator has an inside diameter of .5″ and a length of 15″. We have **80 psi** shop air available, and will assume a **compressibility factor** of 2% per **1000 psi**. From supplier catalogs, we find that the **pre-travel** for available boosters is **1.25″**. Using calculations which assume no friction losses, we need to determine:

1. What hydraulic **pressure** will produce the **force** required?
2. What **intensification factor** (booster ratio) is needed?
3. What is the **total volume** of oil subjected to high pressure?
4. If the pressure is **2000 psi** or more, how much is this **total volume** of oil compressed?
5. For various **ram** diameters, what would be the **total booster stroke**, considering **pre-travel**, **compressibility**, and oil volume needed for the **working** stroke?

6. What size dual pressure booster would be suitable, considering reasonable piston and ram diameters and stroke length? (see Figure 10-8.)

First find the pressure needed on the 3″ actuator to produce 20,000 lbs. of force, using

$$P = F/A:$$

$$A = .785d^2$$
$$A = .785(3)^2$$
$$A = .785(9)$$
$$A = 7.065 \text{ sq. in.}$$
$$P = F/A$$
$$P = 20,000 \div 7.065$$
$$P = 2831 \text{ psi needed.}$$

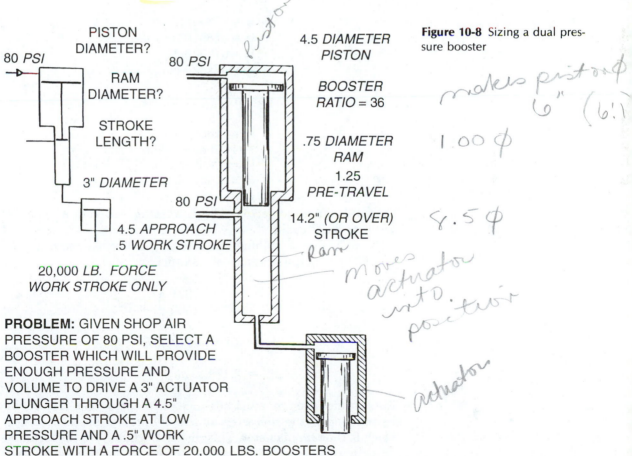

Figure 10-8 Sizing a dual pressure booster

PISTON DIAMETER?

80 *PSI*

RAM DIAMETER?

STROKE LENGTH?

3″ *DIAMETER*

4.5 *APPROACH*
.5 *WORK STROKE*

20,000 *LB. FORCE*
WORK STROKE ONLY

80 *PSI*

80 *PSI*

4.5 *DIAMETER PISTON*

BOOSTER RATIO = 36

.75 *DIAMETER RAM*

1.25 *PRE-TRAVEL*

14.2″ *(OR OVER) STROKE*

makes piston⌀ 6″ (6:1

1.00 ⌀

8.5 ⌀

piston

Ram

moves actuator into position

actuator

PROBLEM: GIVEN SHOP AIR PRESSURE OF 80 PSI, SELECT A BOOSTER WHICH WILL PROVIDE ENOUGH PRESSURE AND VOLUME TO DRIVE A 3″ ACTUATOR PLUNGER THROUGH A 4.5″ APPROACH STROKE AT LOW PRESSURE AND A .5″ WORK STROKE WITH A FORCE OF 20,000 LBS. BOOSTERS ARE AVAILABLE WITH 1.25″ PRE-TRAVEL.

Now find the intensification factor, or **booster ratio** needed to increase the 80 psi shop air to 2831 psi:

$$\text{Booster Ratio} = \frac{\text{Output Pressure}}{\text{Input Pressure}} = \frac{2831}{80}$$

Booster Ratio = 35.4, or 36.

Assume that the booster ram reaches the end of its stroke just as the actuator completes the **work** stroke. This represents a "worst case" situation, because otherwise there is **more** than enough oil available. The oil subjected to high pressure consists of the volume displaced during the **total** 5″ stroke of the 3″ diameter actuator **plus** the oil in the 15″ line. We use the formula for the volume of a cylinder, letting h represent the actuator stroke and line length. In the actuator:

$$V = .785d^2h$$
$$V = .785(3)^2(5)$$
$$V = .785(9)(5)$$
$$V = 35.325 \text{ cu. in.}$$

In the line:

$$V = .785d^2h$$
$$V = .785(0.5)^2(15)$$
$$V = .785(.25)(15)$$
$$V = 2.944 \text{ cu. in.}$$

The total volume **subjected** to high pressure is therefore 35.325 + 2.944, or **38.27 cu. in.**

The **reduction** in volume, using a compressibility factor of **2% per 1000 psi**, and a pressure of **2831 psi** is:

$$V_{compr} = .02 \left(\frac{2831}{1000} \right) (38.27)$$

$$V_{compr} = .02(2.831)(38.27)$$

$$V_{compr} = 2.17 \text{ cu. in.}$$

The distance the ram must **advance** to compress this 2.17 cu. in. depends upon the **diameter** of the cylindrical shape it takes, which is the **ram** diameter. This distance will be h in the formula:

$$h = \frac{V}{.785d^2} .$$

We cannot determine this, however, until we know how far the ram must advance just to move enough oil for the **work** stroke.

Although the oil in the 15″ line is included in the **compression** calculation, the calculation of the booster ram work stroke includes **only** the oil displaced by the **actuator** work stroke of .5″. We learned earlier that the ratio between actuator and ram stroke **lengths** is the same as the ratio between the **squares** of their diameters. We multiply the **actuator high pressure** work stroke by this **ratio** to find the **ram** work stroke:

$$W_{ram} = \frac{(D_{act})^2}{(D_{ram})^2}(W_{act})$$

where

$$W_{ram} = \text{Work stroke of the ram}$$
$$W_{act} = \text{Work stroke of the actuator}$$
$$D_{ram} = \text{Diameter of the ram}$$
$$D_{act} = \text{Diameter of the actuator.}$$

For a .5″ diameter ram you get:

$$W_{ram} = \frac{3^2}{.5^2}(.5) = \frac{9}{.25}(.5) = 18″.$$

For a .75″ diameter ram you get:

$$W_{ram} = \frac{3^2}{.75^2}(.5) = \frac{9}{.5625}(.5) = 8″.$$

For a 1″ diameter ram you get:

$$W_{ram} = \frac{3^2}{1^2}(.5) = \frac{9}{1}(.5) = 4.5″.$$

The **booster ratio** of **36** requires that the booster **piston** diameter be **6** times the **ram** diameter. Again, the design decision includes **space** considerations and the boosters commercially **available**. Assume we decide to use a booster with a 4.5″ **piston** and a .75″ ram. This provides the booster ratio needed and requires an 8″ ram work stroke. We can now calculate the distance the ram travels in compressing the oil by **2.17** cubic inches. Use the formula for finding the height of a cylinder, with h representing the ram **travel**:

$$h = \frac{V}{.785d^2}$$

$$h = \frac{2.17}{.785(.75)^2}$$

$$h = \frac{2.17}{.785(.5625)}$$

$$h = \frac{2.17}{.4416}$$

$$h = 4.91'' \text{ ram travel.}$$

We now have all three figures needed to determine the total ram travel:

1.25″ pre-travel + 4.91″ compression + 8.0″ ram work stroke = 14.16″ TOTAL.

A reasonable choice of booster would have a 4.5″ piston, .75″ ram, and the next available stroke length over 14.2″.

BOOSTER LIMITATIONS

The resolution of every design problem involves a **series** of decisions. These include the **general** approach to pursue, the **specific** method for accomplishing each objective, the kind of **hardware** and **components** to use, and the specific **products** and their assembly.

In this and previous chapters, we learned of various means by which high force levels can be developed hydraulically. Each has its limitations, however. Space considerations restrict the use of large **diameter** cylinders. **Tandem** cylinders may be a solution where the diameter must be restricted, but **length** is no problem. They are, however, more **expensive** than conventional cylinders. **Stepped-piston** cylinders offer variable **approach** and **work stroke** speed and force, but both are very **limited**. Each proposed solution needs to be evaluated with regard to both its **ability** to accomplish the required task and how it **compares** with other choices available.

Often the choice is a simple matter of **expense**. A booster installation will usually cost more than the other methods. Generally, a booster is used only where a **single** actuator is to be powered at high pressure, or where two or more actuators operate **together**. Most of the advantages are lost if valves, piping, and control devices must be located **between** the booster and actuators. If the task requires a **long** work stroke at **high** pressure, and the cycle must be repeated at **short** intervals, a booster usually offers no advantage.

In many hydraulic systems, functions such as starting, stopping, pressure, flow rate, and valving are initiated by the position or status of components such as valves. In one situation, for example, flow occurs only when **two** valves are open at the **same time**. In another, flow occurs when **either** or **both** valves are open. In still another, flow occurs when one **or** the other is open, but **not** when **both** are open. This is known as **logic control**. One of the most common examples of logic control is the **interlock** circuit used as a safety feature on hydraulic presses such as the one shown in Figure 10-9. It ensures that the operator's hands are clear of the work area when the press is operated.

Federal safety regulations require that manual presses have **two** pushbuttons or levers, which must be pressed or turned at the **same time**. Furthermore, the circuitry must be designed so that both must be **released** before the press will re-cycle. This pre-

INTERLOCK CIRCUIT CONTROL

Figure 10-9 Hydraulic presses are equipped with interlock circuits to comply with federal safety regulations. *Courtesy of Power Team Div., SPX Corp., Owatonna, MN*

Figure 10-10 Two-pushbutton interlock circuit

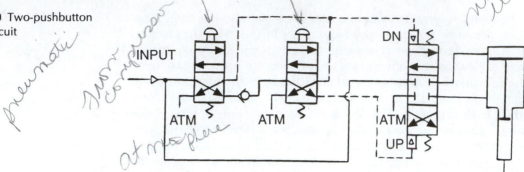

[handwritten annotations: pneumatic; shop compressor; 2 pos 4 way; Push to indx; at mosphere; more return loop]

SCHEMATIC ILLUSTRATION (SHOWN WITH NO INPUT PRESSURE): WHEN INPUT LINE IS PRESSURIZED, THE **UP** PILOT IS ACTUATED TO RETRACT THE BOOSTER RAM.

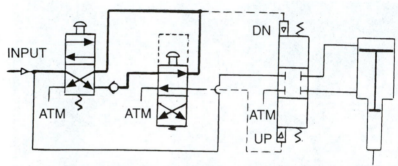

ONE PUSHBUTTON PRESSED: INPUT AIR GOES TO ATMOSPHERE. NO PRESSURE IN EITHER PILOT LINE.

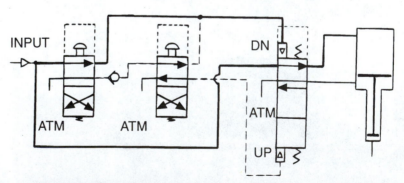

BOTH PUSHBUTTONS PRESSED: DN PILOT IS ACTUATED. BOOSTER RAM EXTENDS.

vents the operator from taping or otherwise holding down one pushbutton permanently to permit one-hand operation.

The interlock circuit is shown schematically in Figure 10-10. With **no pressure** in the input line, the pilot-operated directional control valve is **spring-centered.** All valve ports are **blocked.** This prevents movement of the booster in either direction, which in turn locks the actuator in place. When the input line is **pressurized,** the booster ram will retract **only** if **neither** pushbutton is depressed. As you can see on the illustration, pressure can reach the **UP** pilot of the directional control valve only if **both** pushbuttons are **released.** This prevents an operator from fastening one pushbutton down **permanently** and operating the press with one hand, creating a **safety** hazard.

If only **one** pushbutton valve is operated, air from the input line passes through it and through the **other** pushbutton valve out to the **atmosphere.** There can be no pressure developed to operate the **DN** pilot. When **both** pushbuttons are depressed, the vent lines to atmosphere are blocked. Input line pressure operates the **DN** pilot, shifting the valve. This same input line pressure acts upon the booster piston to extend the ram.

CHAPTER SUMMARY

Hydraulic devices used to increase fluid pressure are called **pressure intensifiers,** or **boosters.** Units are readily available which will multiply pressures up to **50-to-1,** and even greater increases are possible. Booster operation is based upon the relationship between **area, force,** and **pressure.** Pressure intensification results from the difference in **area** between the booster **piston** and the **end** of the **ram.** A booster may be appropriate for any application in which a **small** amount of **high** pressure **liquid** is needed and a **large** amount of **lower** pressure liquid **or** gas is available. Boosters are used to **maintain** high pressure with no loss of **energy.** Since they are **mechanical** in nature, boosters present no **explosion** or **fire** hazards.

With a booster, directional and pressure control are accomplished at **low** pressure while high pressure fluid is **isolated** between the booster and actuator. This permits the use of less expensive **piping** and **components** and **safer** operation. Since the input and output fluids in a booster are kept apart, the two can have different **characteristics.** Boosters are used to **isolate** fluid at **high** pressure in the parts of the system where high pressure is **needed.** This increases operating **economy, safety,** and **efficiency.**

When **shop air** is used to power a booster, certain hydraulic capabilities can be achieved without the need for the piping and components of a conventional hydraulic **system.** In **most** applications, oil pressurized by a booster is directed to an actuator, where it is converted into **force**. In certain **metal-forming** operations, however, air pressure is converted into high pressure **water** which expands thin-walled metal cylinders into desired shapes. **Double-acting** boosters provide a **continuous** supply of high pressure fluid by **alternating** between two output ports. A **dual pressure** booster is used when actuator motion consists of an **approach** stroke which needs only a **low** force followed by a **work** stroke requiring **high** force.

Oil subjected to pressures of **2000 psi** and over is reduced in volume by an amount between **1%** and **3%** per **1000 psi**. This is called the **compressibility factor,** and must be taken into consideration when calculating booster size. The **intensification factor,** or **booster ratio,** is a figure indicating the ratio between **output** and **input** pressures in a booster. The ratio between piston and ram **diameters** is the same as the **square root** of the **booster ratio.** The ratio between actuator and ram **stroke lengths** is the same as the ratio between the **squares** of their diameters.

Finally, in many hydraulic systems, functions are initiated by the **status** of certain components, such as open or closed valves. This is called **logic control.** The **interlock** circuit used as a safety feature on hydraulic presses is an example of logic control.

PROBLEMS

10.1 Why must the high pressure output of a booster be a liquid, rather than air?

10.2 What is the booster ratio of a booster with a 4.5″ diameter piston and a 1.5″ diameter ram? $\frac{4.5^2}{1.5^2} = \frac{20.25}{2.25} = 9:1$

10.3 What is the booster ratio of a booster with a 6.25″ diameter piston and a 1.25″ diameter ram?

10.4 What is the intensification factor of a booster which produces 2400 psi from 80 psi shop air? 2400 psi / 80 psi = 30

10.5 What is the intensification factor of a booster which produces 1430 psi from 65 psi shop air?

10.6 What is the intensification factor of a booster which produces 455 psi from 65 psi shop air? 1430 psi / 65 psi = 22

10.7 What is the intensification factor of a booster which produces 520 psi from 80 psi shop air?

10.8 List three reasons why a booster might be used to increase pressure in a system even when the pump rating is higher than the high pressure developed.

10.9 Describe two advantages of using an air pressure regulator instead of a pressure relief valve to limit booster output pressure.

10.10 If the ram of a booster has a diameter of 1.5″, what piston diameter will result in an intensification factor of 16? $\frac{x^2}{1.5^2} = 16$ 6 = ⌀ of PISTON

10.11 If the ram of a booster has a diameter of 2.5″, what piston diameter will result in an intensification factor of 9?

10.12 If the ram of a booster has a diameter of .75″, what piston diameter will result in an intensification factor of 25?

10.13 If the ram of a booster has a diameter of .5″, what piston diameter will result in an intensification factor of 12.25?

10.14 Using a compression factor of 2% per 1000 psi, calculate the reduction in volume of 45 cu. in. of oil when subjected to 3100 psi.

10.15 Using a compression factor of 2% per 1000 psi, calculate the reduction in volume of 39 cu. in. of oil when subjected to 4800 psi.

10.16 Using a compression factor of 3% per 1000 psi, calculate the reduction in volume of 39 cu. in. of oil when subjected to 12,000 psi.

10.17 Using a compression factor of 3% per 1000 psi, calculate the reduction in volume of 22 cu. in. of oil when subjected to 14,500 psi.

In solving the following problems, assume that boosters are available with these dimensions:

Ram diameter: .5″, 1″, 1.5″, 2″, 2.5″, 3″
Piston diameter: 7″, 7.5″, 8″, 8.5″, 9″, 9.5″, 10″
Stroke: 7″, 7.5″, 8″, 9″, 10″, 12″, 14″, 16″

The dual pressure boosters have a 1.5″ pre-travel.

10.18 You have an application in which a 3.5″ diameter plunger must exert 7000 lbs. of force through a 6″ stroke. With an input of 80 psi shop air, what piston, ram, and booster stroke will do the job?

10.19 You have an application in which a 2″ diameter plunger must exert 6000 lbs. of force through a 4″ stroke. With an input of 80 psi shop air, what piston, ram, and booster stroke will do the job?

10.20 You have an application in which a 4″ diameter plunger must exert 12,000 lbs. of force through a 3.8″ stroke. With an input of 80 psi shop air, what piston, ram, and booster stroke will do the job?

10.21 You have an application in which a 3.75″ diameter plunger must exert 14,000 lbs. of force through a 4.2″ stroke. With an input of 80 psi shop air, what piston, ram, and booster stroke will do the job?

10.22 A 2.5″ diameter plunger used in a metal punch must advance a total of 4.5″, consisting of a 4.25″ approach and a .25″ work stroke with a force of 18,000 lbs. The high pressure line between the booster and actuator has an inside diameter of ½″ and a length of 18″. The input is 80 psi shop air. Assume a compressibility factor of 2% per 1000 psi. What piston, ram, and booster stroke will do the job?

10.23 A 3″ diameter plunger used with a forming die must advance a total of 5.5″, consisting of a 4.5″ approach and a 1″ work stroke with a force of 22,000 lbs. The high pressure line between the booster and actuator has an inside diameter of ¾″ and a length of 20″. The input is 80 psi shop air. Assume a compressibility factor of 2% per 1000 psi. What piston, ram, and booster stroke will do the job?

10.24 A 3.5″ diameter plunger used with a forming die must advance a total of 5″, consisting of a 4.25″ approach and a .75″ work stroke with a force of 24,000 lbs. The high pressure line between the booster and actuator has an inside diameter of ⅜″ and a length of 24″. The input is 80 psi shop air. Assume a compressibility factor of 3% per 1000 psi. What piston, ram, and booster stroke will do the job?

10.25 A 2.75″ diameter plunger used with a forming die must advance a total of 5″, consisting of a 3.75″ approach and a 1.25″ work stroke with a force of 40,000 lbs. The high pressure line between the booster and actuator has an inside diameter of ⅝″ and a length of 12″. The input is hydraulic oil at 180 psi. Assume a compressibility factor of 3% per 1000 psi. What piston, ram, and booster stroke will do the job?

System Maintenance

11

In **many** respects, the operation and maintenance of fluid power systems involve more critical elements and potential problem areas than those powered by any other means. The **fluid** itself is a source of problems, by changing **composition** as a result of heat or chemical reaction, by **reacting** with materials used in circuit components and causing **corrosion** and **oxidation**, and by **distributing** harmful substances throughout the system. Even **newly manufactured** parts often contain **particles** and **shavings** which cause malfunctions unless flushed out immediately.

Placement and **alignment** of valves, trip levers, and actuators are often critical. Even a **small** amount of **air**, hidden in the far reaches of a vast network of piping and valves can render a hydraulic system **inoperable**. Deceptively high **forces** on piston rods, and stored **energy** in accumulator circuits can be extremely hazardous to personnel and equipment. Whatever the other considerations, **safety** takes priority over **all** other concerns at every stage in the design, installation, operation, and maintenance of any system. Obviously no one wants to be involved in, or be held responsible for an industrial **accident**. There are, however, other and more **subtle** justifications for maintaining a secure facility. Employees are more **productive** when they feel protected. A pleasant, cooperative **work environment** is reinforced by worker confidence that management **cares** about their physical well-being. Industrial accidents can be **financially** devastating, and more than one company has been **destroyed** for failure to maintain a **safe workplace**.

KEYS TO EFFECTIVE MAINTENANCE

Effective and efficient maintenance of an industrial hydraulic installation does not begin and end with periodic servicing and repair. It requires planning, preparation, and execution of a series of **interrelated** activities. Good **design** devotes as much attention to convenience of **maintenance** as to proper **function**. Proper **training** includes **theory** of system operation, not just "cookbook" procedures. We can envision hydraulic system maintenance as a four-part **integrated** function with **safety** at its core, as symbolized in Figure 11-1.

Figure 11-1 Essential elements of maintenance

Proper Design

System design typically begins with a series of calculations and choices focused on the **function** to be performed. Obviously this is **primary**, but it is only **one** aspect of the task. Once it has been established that the function **can** be performed in a particular way, modifications, additions, components, and procedures must be instituted to ensure that it **will** be performed **safely**. The entire proposed system must then be reexamined from the standpoints of **reliability, inspection, maintenance,** and **repair.** Areas which need to be **checked** or **serviced** routinely should be conveniently placed, to lessen the likelihood of neglect. The overall **layout** must provide for convenient repairs and replacement without requiring major disassembly. A truly competent designer will **mentally** build and service a system while it is still on the CAD screen, to ensure that these details have been addressed.

Component Selection and Installation

Not even the best design will result in satisfactory operation unless all elements of the circuit meet industrial quality standards **and** are installed correctly. New parts are a major contamination source. Careful inspection and cleaning of each part, including those **newly purchased,** is an important step. Components selected for portions of a circuit where failure would create **special** problems should have **higher** ratings than normally needed, to improve **reliability.** Materials used for seals and components must be **compatible** with the hydraulic **fluid** used. All parts must be **installed** in accordance with manufacturers' specifications, then **operationally checked.** After an initial start-up period, it is good practice to **drain** the system and **replace** the fluid. This serves to eliminate contaminants left in new parts after they were manufactured and also those that found their way into the system during assembly.

Trained Personnel

Every aspect of the installation, operation, and maintenance of modern industrial equipment requires the care of knowledgeable people. Proper training goes far beyond simply instructing each worker to perform specific tasks, as was accepted practice for many years. The **trained** employee, knowing **how** and **why** each part of a system works, is more likely to operate it more **efficiently,** produce a better quality **output,** identify and attend to **malfunctions,** and do so more **safely.**

Employee training, whether in the form of a **formal** course in fluid power or a half-day **updating** session, needs to be carefully organized and presented to be fully effective. It is important that people not be given **conflicting** information. **Retraining** must be available in response to **changes** in equipment, policies, procedures, and regulations. Because of the complexity of equipment and the rapid advances in technology, even **recent** graduates of technical schools and colleges can benefit from frequent updating.

Systematic Maintenance Program

A system cannot be expected to perform reliably and efficiently if attended to only in response to a **breakdown.** From the time of initial **start-up,** a system should have a **planned, documented** maintenance program in place. **Specific** elements **vary** among installations, but most programs should include at least these:

1. A prescribed plan for **routine** maintenance such as peri-

odic **inspections, lubrication,** and **replacement** of parts especially vulnerable to wear;

2. Documentation of data such as **pressures, speeds, measurements, timing,** and **flow rates** recorded while the system is operating **properly.** This is used as a reference for troubleshooting malfunctions;

3. A **manual,** listing all the **components** in the system by **model** and **serial number,** with names and addresses of parts **suppliers;** and

4. A permanent **record** of all routine maintenance performed, malfunctions experienced, repairs accomplished, or parts replaced, and any observations of system operation considered significant.

SYSTEMATIC TROUBLESHOOTING

Rare indeed is the operating system, whether electrical, mechanical, pneumatic, or hydraulic which continues to function indefinitely without occasional malfunction. The origins of breakdowns can be traced to every step in design, building, operation, and maintenance. Just as the **training** and **duties** of those who design, maintain, or troubleshoot are different, so must be their way of **thinking.** The **designer** must focus on the proposed **function** to be achieved, and creates a scheme to accomplish it. **Maintenance** responsibility begins with a **working** structure in place, and its task is to attempt to **sustain** proper operation. This calls for a **planned** program involving constant **monitoring** of known problem indicators, systematic **servicing** of parts subject to wear or other changes, and **replacement** of components **before** they fail. Analysis and correction of malfunctions, more commonly known as **troubleshooting,** is in a sense the **reverse** of **design** and is regarded by many as a blend of **art, imagination, training, intuition,** and **experience.** It has been said that **textbooks** are written for **designers,** and **owner's manuals** for **service technicians,** but **troubleshooters** must learn from **handwritten notes, word-of-mouth,** and **hard experience.**

Several factors influence the **approach** to take in troubleshooting a malfunction. If the system is newly designed and installed, it is probably best to begin by checking for **installation** errors or **neglected maintenance.** If **similar** systems exist, and have been in operation for years, then **their** operating histories become a valuable source of information. Once a system has operated for awhile, its **own** record of maintenance and repairs becomes a primary source.

The system's **environment** leads one to suspect certain sources of problems. Hydraulic systems on earth-moving equipment, or aboard ship, are subjected to conditions very different from those mounted on firm foundations in a building. We would expect to encounter more breakdowns from dirt and abrasive grit contamination in a foundry than in a food-processing facility.

Even the very **nature** of the manner in which malfunctions **develop** varies. Sometimes the actual breakdown is the **last** element in a **sequence.** You may need to trace back through a **series** of symptoms to locate the root of the problem. Consider, for example, a cylinder whose operation has become sluggish. **One** possible cause is a lowered fluid viscosity. **One** cause of lowered viscosity is excessive heat. **One** origin of overheating is low fluid level in the tank. In each case, there are **other** possibilities. This is why, in the absence of a **record** of malfunction symptoms and causes, it may be best to verify proper installation and maintenance practices first. Often a system or component tolerates a **single** deficiency and breaks down only after a **second** part fails. Binding in a piston rod may never be noticed until its operating pressure becomes marginal, for example. Eliminating only the **obvious** symptoms leaves the system still exposed to problem development.

Still another challenge for the troubleshooter is the matter of sorting out **causes** and **effects.** In searching for the cause of oil leakage, for example, you might find a damaged bearing seal. Further examination discloses that the rod is badly scratched. Was the seal damaged by the scratched rod, causing the leak, or did a particle of abrasive in the fluid become imbedded in the seal, damaging it and scratching the rod?

Hydraulic systems are far too complex and varied for a simple "cookbook" list of procedures to be meaningful for all. There are, however, some general guidelines for the beginning troubleshooter.

Know the system's characteristics when it works. This involves personal observations backed up by a written record of speeds, flow rates, pressures, times, and related specific data for reference when malfunctions occur.

Following malfunction, gather data before inspecting the system. Conducting a preliminary analysis is easier and more efficient if you begin with the **operator's input.** Were there differences in operation prior to failure? Have certain pressures been changing? Were recent or frequent adjustments necessary? We often find that an operator's analysis of **why** failure occurred may be faulty, but observations of **what** took place are quite reliable.

Take all appropriate safety precautions first. It is impossible to predict what hazardous conditions may have been **left** in place at the time of breakdown. A heavy load may be supported hydraulically, ready to drop should a valve be operated. An accumulator may be pressurized, ready to begin an operating cycle. The rod of a cylinder may be binding, ready to release explosively if disturbed. The unavoidable hazards present in an **operating** system are easily multiplied tenfold when it **malfunctions.**

Verify the operator's report. This step may take several forms. It may begin with the inspection of a faulty **output,** such as parts produced with incorrect dimensions. It may mean comparing certain **pressures** or **control settings** with **previous** records or operating manuals. It may be a matter of attempting to **operate** the machine to see if the malfunction occurs on **every** cycle, or only **occasionally.** Before beginning to **fix** anything, it is important to make sure the problem lies within the system you are **looking** at, and is not the end result of something that occurred **elsewhere,** such as a power interruption.

Perform your inspection systematically. Malfunctions stem from two general sources: (a) a system component **itself** fails, or (b) some form of external **input** is interrupted, or presented incorrectly. Several factors, based largely upon experience and intuition, influence your decision on where to look first. A component which has failed frequently before, or which has lasted long beyond its expected operating life, is of course suspect. Discoloration from overheating is an invitation to closer examination. Particles of material from a seal or bearing are indicators of trouble spots. If a **possible** source of a given problem is very **convenient** to check, it makes sense to check it **first** even if it is the **least** likely. Sometimes you get lucky!

Inspect for contributing causes. The **direct** cause of malfunction is not always the **sole** basis for a breakdown. A loose connection may be the result of excessive force from **binding** elsewhere. Insufficient force on the end of a piston rod may be the result of friction from high heat, **or** a defective pressure relief valve, **or** mechanical binding, **or** air in the system, **or** a leaking seal, **or** a combination of factors. Good designers make allowances for some decline in component performance. System failure results when either a **single** part changes **drastically** or when **two** or more deteriorate **marginally.** A competent troubleshooter considers **both** possibilities.

FLUID MAINTENANCE

More problems can be traced back to improper or negligent maintenance of the **fluid** than to any other single factor in system operation. Maintenance procedures, service schedules, and the degree of filtration needed vary widely. An immersed **strainer** may provide sufficient filtration for a system operating in a clean, temperature-controlled environment and protected by a simple pressure relief valve. At the other extreme, a hydraulically driven precision machine tool in a factory may require an **elaborate** filtering system and a rigorous **maintenance** schedule. **No** system, however, can be expected to function **indefinitely** without regular attention.

The **fluid itself** can be the cause of system breakdown. When the wrong type of fluid is used, or when it is overheated, it can damage material in seals and other parts. Only the **exact fluid specified** for the system should be used. Chemical compositions of fluids vary widely in capabilities and characteristics. Overheating causes oil to lose its lubricating ability and change its chemical makeup. Either is potentially damaging. Heat and water convert hydraulic oil into gum and sludge which block **ports**, increase **friction**, and cause **valve malfunctions**. When contaminants are introduced in, or produced by, any part of a hydraulic circuit, the **fluid** carries these particles or chemicals to areas where they can damage components or initiate malfunctions.

Fluid maintenance begins with **installation,** even before assembly of the individual **components**. Pipe and screw **threads** need to be cleaned and inspected for jagged edges that might break off and be carried in the fluid. The tank should be located away from any **heat** sources and should have sufficient clearance all the way around to allow adequate circulation of **cooling air**. At the bottom, clearance of at least **6″** is generally recommended and of course nothing should be placed against the sides or on top. Improper **storage** practices or careless **filling** procedures can result in contamination of an otherwise clean, functional circuit. It is important that the **filler neck** of the tank be located so that it can be filled conveniently and without letting dirt particles enter. It should be kept covered, if possible, when not being used. The **spouts** of storage cans or drums should also be protected, either by storing the containers on their **sides** or by providing **covers** or both. Avoid **galvanized** containers. Some of the chemicals used as **additives** in hydraulic fluid **dissolve** or **corrode** certain metals, including zinc used in galvanizing.

SEALS

A key factor in keeping a system **leak-free** is the correct selection and installation of **seals,** although that is not the **sole** function of these devices. Seals are used wherever a section of the system may sometime need to be **taken apart** and **reassembled,** or where there is **movement** between two contacting parts. If there is contact but no motion, a **static seal**, or **gasket** is used. **Static seals** are used, for example, wherever a valve, pump, filter, cylinder, or other component is connected to piping. Leakage or contamination at the interface between **moving** parts is deterred by a **dynamic seal**. **Dynamic seals** are found, for example, on pistons, piston rod bearings, and valves.

The most common seal is the simple **O-ring,** used in both static

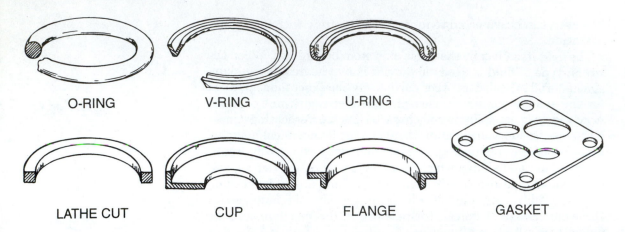

O-RING V-RING U-RING

LATHE CUT CUP FLANGE GASKET

Figure 11-2 Seal shapes

and dynamic applications. These are available in a variety of materials and standard sizes. **Lathe-cut** rings are similar, but less expensive and are suitable only for **static** sealing. They are produced by slicing synthetic rubber sleeves, similar to flexible piping. When a ring-type seal is subjected to pressure in only one direction, **V-ring** or **U-ring** seals may be used, providing tighter closure. Figure 11-2 illustrates a few of the more common types of seals. Custom-designed **gaskets** are used where static sealing is needed on a complex or multiported surface.

Seals are used to prevent **contamination** and **leakage** of hydraulic fluid. The consequences of **contamination** were described in the section on fluid maintenance. **Leakage** creates another set of problems:

1. Loss of fluid represents needless **expense;**
2. Spilled fluid is a **safety hazard;**
3. **Low** fluid **level** resulting from leakage is often the root cause of **system malfunction;**
4. Leaked fluid can **contaminate** whatever is being **produced** by the system, especially food products; and
5. **Internal** leakage reduces system **efficiency,** and if extreme, can initiate a **malfunction.**

The **selection** or **design** of a seal **material** and **shape** for a specific position in a circuit requires careful consideration of **three** factors. The first is the **purpose** for sealing. A **static** seal must be slightly **compressible** but need not be wear-resistant or made of low-friction material, as is the case with a **dynamic** seal. Where several openings on a flat surface must be sealed, either a single gasket or individual O-rings around the holes may be appropriate. The dynamic seal used on a **reciprocating** shaft is different from one used where a shaft **rotates.** A seal used to **confine fluid** under high pressure differs greatly from one installed just to keep out **contaminants.** The second factor concerns the **operating conditions,** with respect to such variables as pressure, temperature, contaminants, operating cycle, and vibration. A seal which shows ideal characteristics at 200°F may be made of a material which could not be used where temperatures reach 500°F. Dust and grit destroy some materials more quickly than others. Shapes such as V-rings and U-rings have advantages when fluid pressure is from one side only. The final factor to be considered is the **fluid** used. Hydraulic systems use a wide variety of fluid compositions. It is extremely important for the fluid to be **compatible** with any material it contacts, including that of the seals.

PROBLEM DIAGNOSIS

When a **newly installed** hydraulic system fails to perform as expected, diagnosis involves checking many more factors than is the case with an **operating** system that breaks down. The specific functional **requirements** of the system may have been incorrectly **communicated** to the **designer,** or the design **concept** may have errors. Inappropriate **components** may have been chosen, or unauthorized **substitutions** made. **Installation** errors may have altered pressures or flow rates, or made it impossible for valves to operate in proper sequence. The **operator** may not have understood or followed key instructions. Diagnosis of problems in initial start-up often require consultation with the designer or a specialist familiar with a similar system. When a system which **has** operated properly breaks down, however, the range of causal factors is **narrowed** considerably. Now the problem is to determine what has **changed.**

Placement of **pressure gauges** at key locations in a system at the time of initial assembly is a valuable aid both in establishing normal operational characteristics and in future troubleshooting. Each gauge is installed with a **tee** fitting in the working line. A **manual** valve is installed between the fitting and gauge. The valve is normally kept **closed** to protect the gauge from pressure surges and vibrations, and opened only when necessary to take readings. One of the most common indications of a developing problem is slow or erratic actuator movement. Several causes are possible, but lacking other clues it might be best to begin by checking fluid **flow** and **pressure** in the working line.

Figure 11-3 shows the installation of a pressure gauge between the **pump** and **pressure relief valve** for this purpose. First, open the manual valve to the pressure gauge and check system pressure. If lower than normal, the relief valve may be either faulty or incorrectly set. Normal pressure does not, however, ensure that the pump is operating properly. This is determined by measuring

Figure 11-3 Pump test

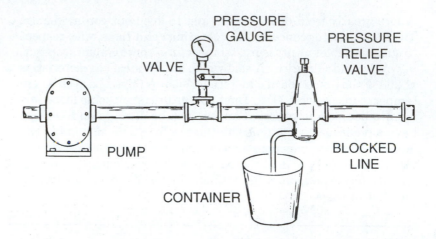

the **flow rate** at **working pressure. Block** flow **downstream** of the pressure relief valve, either by closing a valve or by disconnecting and capping the line. **Lower** the pressure setting of the relief valve by about **25%** and start the pump. Allow the fluid temperature to rise to the normal operating range while **all** the fluid passes through the **relief valve** back to the tank.

Next, collect the full flow from the valve in a container of known capacity, such as a quart or half-gallon measuring can. Disconnect the tank line if necessary. With the **lowered** pressure setting, record the time it takes to fill the container, and calculate the flow rate in GPM. This figure should closely match the pump **rating**. If it does **not**, check the tank fluid level and strainer. The fluid level should normally be at least **3″** above the strainer to avoid creating a **vortex** when pumping. This is similar to the "hole" you see in the water as a sink or bathtub drain. In a hydraulic system, it allows unwanted **air** to get into the oil. Remove and clean the strainer, even though no particles are **visible**. This may be done by blowing pressurized **air** in the **reverse** direction of fluid flow, or by **flushing** it with a fluid compatible with that used in the system.

If the pump output is normal at the **lowered** pressure, excessive **slippage** at working pressure, perhaps from wear, may be the problem. Adjust the pressure relief valve for **normal** working pressure, and again fill the container and calculate flow rate. This should be the same as indicated by the pump rating. If **so**, then further diagnosis should focus on components **downstream**. If **not**, then the source of the malfunction is likely to lie within the **pump** or **relief valve.** System testers are commercially available which provide a convenient means for taking pressure and flow measurements at several locations. Assembled as portable units, they can save considerable time and money in facilities where fluid power is used extensively. An example of a system tester is shown in Figure 11-4.

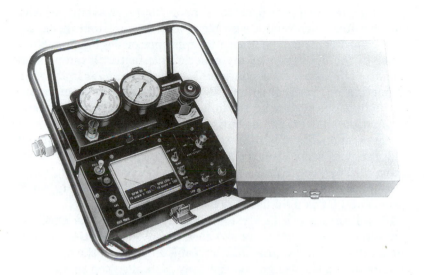

Figure 11-4 Hydraulic system tester. *Courtesy of Power Team Div., SPX Corp., Owatonna, MN*

This is but a brief description of one troubleshooting procedure, focusing on a single type of malfunction. Becoming truly proficient requires first learning hydraulic principles, then acquiring work experience solving problems in a variety of situations under skilled supervision. You have, in mastering this text, already taken the first step.

CHAPTER SUMMARY

Maintenance of a fluid power system is in many respects more challenging than with other types. Whatever the other problems and considerations, however, **safety** is most important.

Effective maintenance consists of a series of **interrelated** activities, including proper **design**, quality **components**, trained **personnel**, and a **systematic program** of routine maintenance, operating data documentation, part and component inventory, and record-keeping. Central to all these are **safety** considerations and practices.

Just as the **functions** of those who design, maintain and trouble-shoot are different, so must be their way of **thinking**. The **approach** needed to troubleshoot a particular malfunction depends upon several factors. These include whether the system is **new** or has an operational **history**, the **environment** of the system, and the way the malfunction **developed**. Sometimes a system continues to function with a **single** deficiency, but breaks down when a **second** part fails. Often it is difficult to determine by observation which factors are **causes** of problems and which are **effects**, or **results**. The general troubleshooting guidelines are: (1) Know system characteristics when it **works**; (2) When a malfunction occurs, gather data **before** inspecting the system; (3) Take **safety** precautions **first**; (4) **Verify** the operator's report; (5) Work **systematically**; and (6) After diagnosing the problem, look for **contributing** causes.

More problems can be traced to improper or negligent maintenance of the **fluid** than to any other single factor. Examples are the use of **improper** fluid, **contaminated** fluid, **changes** in fluid due to heating or chemical action, and **distribution** of contaminants throughout the system by the fluid. Seals are used to prevent **contamination** and **leakage** of hydraulic fluid. **Static seals** are used where components may need to be taken apart and reassembled, and **dynamic seals** are used to close gaps between parts which **move** with respect to each other. Seal selection or design involves three factors: (1) the **purpose** of the seal, (2) the **operating conditions,** and (3) the **fluid** used.

Troubleshooting a **newly installed** system involves checking many more factors than is the case when an **operating** system breaks down. The designer may have misunderstood the functional **requirements,** the design **concept** may be faulty, the **components** may be inappropriate, or errors may have been made during **installation.** When a system has been in operation for a period of time, the task of the troubleshooter is to determine what has **changed.** Placement of **pressure gauges** at key locations is a valuable aid, both in establishing **normal** system characteristics and in future **troubleshooting.**

PROBLEMS

11.1 Describe briefly three ways in which hydraulic fluid can be a source of maintenance problems.

11.2 How can new parts lead to maintenance problems, and what can be done to avoid these problems?

11.3 List three reasons, other than the obvious fact that no one wants to be injured, that safety is the single most important concern in every aspect of system operation.

11.4 Explain briefly four ways in which job performance improves when workers receive proper training.

11.5 List and explain briefly four elements which should be included in a hydraulic system maintenance program.

11.6 Describe briefly how the functions and responsibilities of a designer, a maintenance worker, and a troubleshooter are different.

11.7 Describe briefly six steps, or guidelines, to follow when troubleshooting a system malfunction.

11.8 List four procedures or precautions which can be taken to prevent hydraulic fluid from causing system malfunctions.

11.9 Why should galvanized containers not be used for storage of hydraulic fluid?

11.10 List five problems which can result from leakage of hydraulic fluid.

11.11 Explain briefly the difference between the functions of static seals and dynamic seals.

11.12 What three factors must be considered when selecting or designing a seal for a specific application?

11.13 Describe four possible origins of problems in a newly installed system which are less likely in one that has been in operation for some time.

11.14 Describe two ways in which dirt and grit particles can be cleaned out of a strainer.

Answers to Even-Numbered Computational Problems

APPENDIX

A

Chapter 1: Learning about Hydraulics

1.14 $n = y+r+b$
1.16 (a) 60 lbs.
      (b) 150 lbs.
1.18 (a) 13.5 gal.
      (b) 90 gal.

Chapter 2: Area, Force, and Pressure

2.2   10.5 sq. in.
2.4   20 sq. in.
2.6   0.56 sq. in.
2.8   22.56 sq. in.
2.10 21.06 sq. in.
2.12 50.24 sq. in.
2.14 200 psi
2.16 840 psi
2.18 21.23 psi
2.20 300 psi
2.22 500 lbs.
2.24 476.4 lbs.
2.26 1413 lbs.
2.28 4 sq. in.
2.30 30 sq. in.
2.32 (a) 82.92.5 lbs.
      (b) 8095 lbs.
2.34 (a) 98.16 lbs.
      (b) 94.24 lbs.
2.36 2.28″ dia.
2.38 2.73″ dia.

Chapter 2: Area, Force, and Pressure (continued)

2.40 2160 lbs.
2.42 4200 lbs.
2.44 16
2.46 9
2.48 3920 lbs.
2.50 537.9 lbs.

Chapter 3: Volume, Capacity, and Fluid Flow

3.4   148.5 cu. in.
3.6   2046 cu. in.
3.8   43.96 cu. in.
3.10 9.65 cu. in.
3.12 415.8 cu. in.
3.14 173.25 cu. in.
3.16 9.77 in.
3.18 14.44 in.
3.20 12.72 in.
3.22 31.23 in.
3.24 10.46 in.
3.26 2.94 in.
3.28 8.69 in.
3.30 20.6 in./min.
3.32 13.93 in./min.
3.34 22.24 in./min.
3.36 19.23 in./min.
3.38 9.44 sec.
3.40 4.94 sec.

Chapter 5: System Calculations

5.2	95%
5.4	89%
5.6	96%
5.8	91%
5.10	74%
5.12	80%
5.14	2.03 HP
5.16	3.94 HP
5.18	7.5 HP motor
5.20	9.5 HP motor
5.22	1.7 HP
5.24	0.55 HP
5.26	1.6 HP
5.28	1.42 HP
5.30	(a) 9537.8 lbs.
	(b) 5.91 HP
5.32	(a) 11,775 lbs.
	(b) 10.7 HP
5.34	2.33 HP
5.36	2.8 HP
5.38	883.5 BTU/hr
5.40	650.4 BTU/hr
5.42	4142 BTU/hr
5.44	1091.35 BTU/hr
5.46	(a) .54″ dia.
	(b) .31″ dia.
5.48	(a) .86″ dia.
	(b) .49″ dia.

Chapter 6: Pumps

6.28	88%
6.30	82%

Chapter 8: Cylinders

8.8	3.99″ dia. piston, 1.107″ dia. plunger
8.10	5.05″ dia. piston, 1.49″ dia. plunger

Chapter 9: Accumulators

9.8	(a) 86.55 cu. in.
	(b) 228.78 psi
	(c) 57.68 cu. in.
	(d) 280.71 lbs.
9.10	(a) 84.78 cu. in.
	(b) 283.09 psi
	(c) 36.66 cu. in.
	(d) 500 lbs.
9.12	2 horsepower
9.28	256 psig
9.30	569.23 cu. in.
9.32	266.36 cu. in.
9.34	911.05 cu. in.
9.36	1234.24 cu. in.
9.38	169 cu. in.

Chapter 10: Pressure Intensifiers

10.2	3
10.4	30
10.6	7
10.10	6″ dia.
10.12	3.75″ dia.
10.14	2.79 cu. in.
10.16	14.04 cu. in.

APPENDIX B

English–Metric Conversions

Inch	Inch	Millimeters	Inch	Inch	Millimeters
1/32	.03125	0.794	17/32	.53125	13.494
1/16	.0625	1.588	9/16	.5625	14.288
3/32	.09375	2.381	19/32	.59375	15.081
1/8	**.1250**	**3.175**	**5/8**	**.6250**	**15.875**
5/32	.15625	3.969	21/32	.65625	16.669
3/16	.1875	4.763	11/16	.6875	17.463
7/32	.21875	5.556	23/32	.71875	18.256
1/4	**.2500**	**6.350**	**3/4**	**.7500**	**19.050**
9/32	.28125	7.144	25/32	.78125	19.844
5/16	.3125	7.938	13/16	.8125	20.638
11/32	.34375	8.731	27/32	.84375	21.431
3/8	**.3750**	**9.525**	**7/8**	**.8750**	**22.225**
13/32	.40625	10.319	29/32	.90625	23.019
7/16	.4375	11.113	15/16	.9375	23.813
15/32	.46875	11.906	31/32	.96875	24.606
1/2	**.5000**	**12.700**	**1**	**1.0000**	**25.400**

Multiply	By	To Find	Multiply	By	To Find
BTU/min.	.02356	HP	HP	42.436	BTU/min.
cu. in.	.00433	gallons	gallons	231	cu. in.
feet	.3048	meters	meters	3.2808	feet
gallons	3.785	liters	liters	.2642	gallons
HP	746	watts	watts	.00134	HP
inches	25.400	mm	mm	.03937	inches

Hydraulic Symbols

The symbols shown here are representative of those approved by the American National Standards Institute for use on fluid power diagrams. For the complete standard, write to either the American Society of Mechanical Engineers, United Engineering Center, 345 East 47th St., New York, N.Y. 10017, or the National Fluid Power Association, 3333 N. Mayfair Rd., Milwaukee, Wis. 53222. Ask for document ANSI Y32.10.

LINES

MAIN HYDRAULIC
LINE OR SHAFT

PILOT LINE
(CONTROL)

HYDRAULIC
DRAIN LINE

ENCLOSURE FOR
COMPONENTS

FIXED   *VARIABLE*

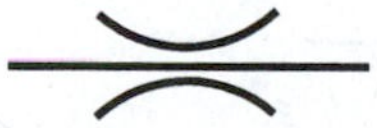

RESTRICTION IN LINE FOR
FLOW CONTROL

HYDRAULIC PUMPS

FIXED
DISPLACEMENT

VARIABLE
DISPLACEMENT
NON-COMPENSATED

VARIABLE DISPLACEMENT
PRESSURE-COMPENSATED

RESERVOIRS (TANKS)

VENTED

PRESSURIZED

LINE BELOW
FLUID LEVEL

LINE ABOVE
FLUID LEVEL

STRAINER BELOW
FLUID LEVEL

STRAINER IN
MAIN LINE

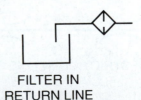

FILTER IN
RETURN LINE

ACCUMULATORS

SPRING
LOADED

GAS
CHARGED

WEIGHTED

PRESSURE INTENSIFIERS (BOOSTERS)

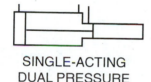
SINGLE-ACTING
DUAL PRESSURE

DOUBLE-ACTING

CYLINDERS

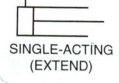
SINGLE-ACTING
(EXTEND)

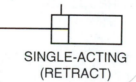
SINGLE-ACTING
(RETRACT)

SINGLE-ACTING
(SPRING RETURN)

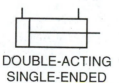
DOUBLE-ACTING
SINGLE-ENDED

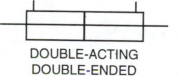
DOUBLE-ACTING
DOUBLE-ENDED

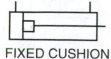
FIXED CUSHION
(ADVANCE ONLY)

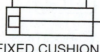
FIXED CUSHION
(RETURN ONLY)

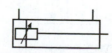
ADJUSTABLE CUSHION
BOTH DIRECTIONS)

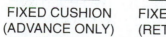

FLOW CONTROL VALVES

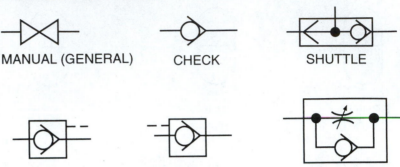

MANUAL (GENERAL)   CHECK   SHUTTLE

CHECK WITH
PILOT TO OPEN

CHECK WITH
PILOT TO CLOSE

ADJUSTABLE
FLOW WITH
BYPASS

PRESSURE CONTROL VALVE

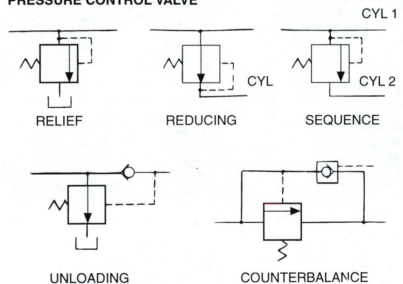

RELIEF   REDUCING   SEQUENCE

UNLOADING   COUNTERBALANCE

DIRECTIONAL CONTROL VALVES

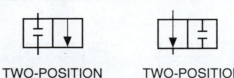

TWO-POSITION
TWO-WAY
NORMALLY CLOSED

TWO-POSITION
TWO-WAY
NORMALLY OPEN

TWO-POSITION
THREE-WAY

DIRECTIONAL CONTROL VALVES (continued)

TWO-POSITION
FOUR-WAY

THREE-POSITION
FOUR-WAY
CLOSED CENTER

THREE-POSITION
FOUR-WAY
OPEN CENTER

EXAMPLES OF OTHER
CENTER CONDITIONS

VALVE ACTUATORS

MANUAL

PUSHBUTTON

LEVER

HYDRAULIC
PILOT

AIR PILOT
PRESSURE-
ACTUATED

AIR PILOT
EXHAUST-
ACTUATED

SOLENOID

FOOT OPERATED
(PEDAL OR
TREADLE)

MECHANICALLY
ACTUATED

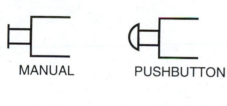

SPRING-CENTERED

SPRING RETURN

Steel Pipe Sizes

For **suction** lines, use Schedule 40 pipe large enough to maintain a flow rate between 2 and 4 fps at pump-rated GPM. For **return** lines, use Schedule 40 pipe large enough to maintain a flow rate between 10 and 15 fps at pump-rated GPM. For **working** lines, use either size, depending on system pressure, large enough to maintain a flow rate between 15 and 20 fps. (For higher pressures, consult a table for Schedule 160 pipe.)

Schedule 40 (Standard Weight)

Nominal Size	Outside Dia.	Inside Dia.	Working PSI*	GPM flow @ 4 fps	GPM flow @15 fps
⅛	.405	.269	2238	0.72	2.70
¼	.540	.364	2173	1.20	4.50
⅜	.675	.493	1797	2.40	9.00
½	.840	.622	1730	3.80	14.0
¾	1.050	.824	1435	6.60	25.0
1	1.315	1.049	1348	11.0	41.0
1¼	1.660	1.380	1124	19.0	70.0
1½	1.900	1.610	1017	26.0	95.0
2	2.375	2.067	864	42.0	156.0
2½	2.875	2.469	941	60.0	222.0
3	3.500	3.068	823	92.0	345.0

*Working pressure calculated using a safety factor of 6. (Burst pressure = 6 × Working pressure)

Schedule 80 (Extra Strong)

Nominal Size	Outside Dia.	Inside Dia.	Working PSI*	GPM flow @ 4 fps	GPM flow @ 15 fps
⅛	.405	.215	3128	0.48	1.80
¼	.540	.302	2938	0.96	3.60
⅜	.675	.423	2489	1.68	5.30
½	.840	.546	2333	2.88	11.0
¾	1.050	.742	1955	6.00	22.0
1	1.315	.957	1815	8.88	33.3
1¼	1.660	1.278	1534	13.6	60.0
1½	1.900	1.500	1403	22.0	83.0
2	2.375	1.939	1224	37.0	138.0
2½	2.875	2.323	1280	53.0	198.0
3	3.500	2.900	1143	82.0	310.0

*Working pressure calculated using a safety factor of 6. (Burst pressure = 6 × Working pressure)

Pipe Fittings

COUPLING

REDUCER

HEX PLUG

CAP

UNION

HEX BUSHING

90° ELL

TEE

LATERAL

STREET ELL

CROSS

45° ELL

Glossary

ABSOLUTE PRESSURE. Fluid pressure as referenced to a perfect vacuum rather than to the pressure of the atmosphere.

ABSORPTION. Process of contaminant removal in which particles are trapped and held within the porous material of a filter.

ADSORPTION. Process of contaminant removal in which particles are blocked from passage through the fine mesh screen of a strainer.

ACCUMULATOR. A container for storing energy in the form of a pressurized fluid.

ACTUATOR. A device which converts fluid power into mechanical force and motion, such as a hydraulic cylinder or motor.

ADDITIVE. A chemical put into a fluid which changes its capabilities, such as a rust inhibitor.

ATMOSPHERIC PRESSURE. Air pressure resulting from the weight of gases above the earth. It varies with elevation above or below sea level, and with weather conditions.

AUTO-IGNITION TEMPERATURE. Level of heat required to cause a sample of oil to catch fire by itself.

BOOSTER. A hydraulic device which raises low pressure in liquid or gas to high pressure in a liquid. Also called an intensifier, or pressure intensifier.

CAP. Cylinder end cover which completely covers the bore, with no opening for a piston rod. Also called the rear end, rear head, back end, blind end, or blind head.

CAVITATION. The presence of potentially damaging cavities, or bubbles, in a stream of liquid.

COMPRESSIBILITY. The tendency of a substance, such as a fluid, to change in volume as a result of pressure change.

CONTAMINANT. Potentially harmful material in a fluid, such as abrasive particles or corrosive chemicals.

CRACKING PRESSURE. The pressure at which a simple pressure control valve begins to allow fluid flow.

CUSHION. Device used in a cylinder to slow piston travel at the end of the stroke by limiting output fluid flow.

CYCLE. A single complete sequence of operations which perform a function, ending with components in their original positions.

FIRE POINT. Temperature at which a given sample of oil, if touched with a flame, will catch fire and stay lit for at least 5 seconds.

FLASH POINT. Temperature at which a given sample of oil will light briefly if touched with a flame.

FLUID. A substance which exists in the form of either a liquid or a gas.

FLUID POWER. Field of applied engineering which deals with the flow and pressure of liquids and gases.

HEAD. Cylinder end cover which has an opening for a piston rod. Also called the rod end, rod head, front end, front face, or front head.

HYDRAULICS. Field of engineering science covering flow and pressure of liquids.

INTENSIFIER. A hydraulic device which raises low pressure in liquid or gas to high pressure in a liquid. Also called a booster.

LAMINAR FLOW. Fluid motion in which all particles move in parallel layers.

MANIFOLD. A fluid conductor consisting of two or more ports or components assembled together as a unit.

MICRON. A millionth of a meter (.00037″).

PILOT. A device or conductor whose function is to control a hydraulic component.

PNEUMATICS. Field of engineering science covering the flow and pressure of gases.

PORT. A plumbing connection by which fluid may enter or leave a hydraulic component, such as a valve or cylinder.

POUR POINT. The lowest temperature at which a given oil will flow enough to be useful, as determined under laboratory conditions specified by ASTM D97–57.

PRESSURE. A measure of force distributed over a specified unit of area.

PRESSURE, GAUGE. Fluid pressure as referenced to the pressure of the atmosphere.

PUMP. A hydraulic device which converts mechanical force and motion into fluid power by causing fluid to flow.

SCHEMATIC. A type of drawing in which parts, components, and connectors are shown in order of their functions rather than in actual physical location with respect to each other.

SEAL. Device or material, such as an O-ring, which prevents or controls flow or seepage of fluid or other material.

SURGE. A sudden, temporary rise in pressure in a fluid.

SYSTEM PRESSURE. The primary pressure setting at which a hydraulic circuit is designed to operate. This pressure may be raised or lowered in specific parts of the system as needed.

TURBULENT FLOW. Fluid motion in which particles move in random directions, creating resistance to movement.

VALVE. A device used in fluid power to control flow direction, flow rate, or pressure.

VISCOSITY. A measure of an oil's resistance to flow, determined by measuring the time required for a specific amount of oil at a specific temperature to flow through a small opening.

VISCOSITY INDEX. A measure of the change in viscosity of oil as a result of heating, as determined under laboratory conditions specified by ASTM D567–53.

Index